Pollution

Pollution

Edited by

Robert S. Leisner

and

Edward J. Kormondy

American Institute of Biological Sciences

WM. C. BROWN COMPANY PUBLISHERS
Dubuque, Iowa

BIOLOGY SERIES
Consulting Editor
E. Peter Volpe

Foundations for Today
A joint publishing venture between
Wm. C. Brown Company Publishers/BioScience

Copyright © 1971 by Wm. C. Brown Company Publishers

Library of Congress Catalog Card Number: 76—160826

ISBN 0—697—04668—0

All rights reserved. No part of this publication may be reproduced,
stored in a retrieval system, or transmitted, in any form or by any
means, electronic, mechanical, photocopying, recording, or otherwise,
without the prior written permission of the copyright owner.

Second Printing, 1971

Printed in the United States of America

Contents

The Effects of Pesticides 1
William A. Niering

The Sea-Level Panama Canal: Potential Biological Catastrophe 8
John C. Briggs

Thermal Addition: One Step from Thermal Pollution 13
Sharon Friedman

DDT on Trial in Wisconsin 15
Bruce Ingersoll

Who Needs DDT? 17
Thomas H. Jukes

DDT on Trial in Wisconsin—Part II .19
Bruce Ingersoll

DDT Roundup 21

DDT Goes to Trial in Wisconsin 22
Charles F. Wurster

Radioactivity and Fallout: The Model Pollution 28
George M. Woodwell

Pollution—Is There a Solution? 32
John G. New and J. Gary Holway

Monitoring Pesticide Pollution 33
Philip A. Butler

Interactions 37
M.R. Zavon

Some Effects of Air Pollution on Our Environment 42
Vincent J. Schaefer

Technology Assessment 45
Ann C. Barker and Jo Ann Fowler

Water Pollution 48
Robert D. Hennigan

Population Pollution 52
   Francis S.L. Williamson

Research and Development for Better Solid Waste Management 57
   Andrew W. Breidenbach and Richard W. Eldredge

Thermal Pollution 63
   LaMont C. Cole

Detergent Enzymes: Biodegradation and Environmental Acceptability 68
   R.D. Swisher

Oil Pollution 70
   Klaus Rutzler and Wolfgang Sterrer

Toward Safer Use of Pesticides 74
   Sheila A. Moats and William A. Moats

Pesticides: Eggshell Thinning and Lowered Production of Young in Prairie Falcons 81
   J.H. Enderson and D.D. Berger

Poisoning with DDT: Second- and Third-Year Reproductive Performance of *Artemia* 84
   Daniel S. Grosch

Introduction

"Ecology," "Population and Food," "Pollution" are timely topics because of their significance in man's quest for survival. Little wonder then that these titles were selected for the first three collections of articles which have appeared in *BioScience* since January 1968. In responding to the critical issues of our times, *BioScience*, the official publication of the American Institute of Biological Sciences, has consistently devoted a considerable portion of its pages to environmental matters. Of the many articles of high quality bearing a significant environmental message which have been published in the last three years, eight have been selected for these first three anthologies.

Why an anthology when *BioScience* is already available in most college and university libraries, as well as in many high schools? A collection of readings provides a focus for a reader on a single topic, isolating that topic from other articles that appear in a given journal and providing thereby a more comprehensive feel for the particular issue in its various dimensions. More importantly, it provides an inexpensive sourcebook for the many courses, seminars and study groups which are currently dealing with the topic: to attempt to have fifty, or even ten people trying to read a given article from one library copy is certainly foolhardy; to reproduce copyrighted articles with the convenience of xerography borders on illegality. Many of us have been gravely concerned that scientists speak too much and too often to their own kind on critical contemporary problems: an anthology at least allows the potential for reaching that other audience, the lay public.

Doubtless as concern continues in the matters of population and the environment, there will be more articles of value for yet further anthologies. As anxiety mounts over such other issues as control of human evolution, biological effects of chemotherapy as well as of "drugs," surgical engineering by organ transplant or the use of prosthetics, the pages of *BioScience* will certainly reflect these problems and thus suggest the compilation of articles into other anthologies at given intervals. Time will tell.

The Effects of Pesticides

William A. Niering

The sound use of pesticides must be evaluated on the basis of their effects on the total environment. Persistent insecticides adversely affect nontarget organisms, accumulate in food chains, appear to lower reproductive potential and may have delayed genetic responses. Indiscriminate use of herbicides result in unnecessary habitat destruction. Their sound ecological use result in high conservation and aesthetic values. (BioScience 9, no. 9, p. 869-875)

The dramatic appearance of Rachel Carson's *Silent Spring* (1962) awakened a nation to the deleterious effects of pesticides. Our technology had surged ahead of us. We had lost our perspective on just how ruthlessly man can treat his environment and still survive. He was killing pesty insects by the trillions, but he was also poisoning natural ecosystems all around him. It was Miss Carson's mission to arrest this detrimental use of our technological achievements. As one might have expected, she was criticized by special vested industrial interests and, to some degree, by certain agricultural specialists concerned with only one aspect of our total environment. However, there was no criticism, only praise, from the nation's ecosystematically oriented biologists. For those who found *Silent Spring* too dramatic an approach to the problem, the gap was filled two years later by *Pesticides and the Living Landscape* (1964) in which Rudd further documented Miss Carson's thesis but in more academic style.

The aim of this chapter is to summarize some of the effects of two pesticides — insecticides and herbicides — on our total environment, and to point up research and other educational opportunities for students of environmental science. The insecticide review will be based on representative studies from the literature, whereas the herbicide review will represent primarily the results of the author's research and experience in the Connecticut Arboretum at Connecticut College. Although some consider this subject controversial, there is really no controversy in the mind of the author—the issue merely involves the sound ecological use of pesticides only where necessary and without drastically contaminating or upsetting the dynamic equilibrium of our natural ecosystems. I shall not consider the specific physiological effects of pesticides, but rather their effects on the total environment — plants, animals, soil, climate, man — the biotic and abiotic aspects.

Environmental science or ecosystematic thinking should attempt to coordinate and integrate all aspects of the environment. Although ecosystems may be managed, they must also remain in a relative balance or dynamic equilibrium, analogous to a spider's web, where each strand is intimately interrelated and interdependent upon every other strand.

The Impact of Insecticides

Ecologists have long been aware that simplifying the environment to only a few species can precipitate a catastrophe. Our highly mechanized agricultural operations, dominated by extensive acreages of one crop, encourage large numbers of insect pests. As insurance against insect damage, vast quantities of insecticides are applied with little regard for what happens to the chemical once it is on the land. Prior to World War II, most of our insecticides were nonpersistent organics found in the natural environment. For example, the pyrethrins were derived from dried chysanthemum flowers, nicotine sulphate from tobacco, and rotenone from the tropical derris plants. However, research during World War II and thereafter resulted in a number of potent persistent chlorinated hydrocarbons (DDT, dieldrin, endrin, lindane, chlordane, heptachor and others) to fight the ever-increasing hordes of insects, now some 3000 species plaguing man in North America.

In 1964, industries in the United States produced 783 million lb. of pesticides, half insecticides and the other half herbicides, fungicides, and rodenticides. The application of these chemicals on the nation's landscape[1] has now reached the point where one out of every ten acres is being sprayed with an average of 4 lb. per acre (Anonymous, 1966).

Positive Effects on Target Organisms

That market yields and quality are increased by agricultural spraying ap-

---

This paper will appear in the symposium volume *Environmental Problems* to be published by J. B. Lippincott, Summer, 1968.

The author is Professor of Botany and Director of the Connecticut Arboretum at Connecticut College.

[1] Dr. George Woodwell estimates that there are 1 billion lbs. of DDT now circulating in the biosphere.

pears to have been well documented. Data from the National Agricultural Chemical Association show net increased yields resulting in from $5.00 to $100.00 net gains per acre on such crops as barley, tomatoes, sugar beets, pea seed, and cotton seed. However, Rudd (1964) questions the validity of these figures, since there is no explanation just how they were derived. His personal observations on the rice crop affected by the rice leaf miner outbreak in California are especially pertinent. The insect damage was reported as ruining 10% to 20% of the crop. He found this to be correct for some fields, but most of the fields were not damaged at all. In this situation, the facts were incorrect concerning the pest damage. It appears that not infrequently repeated spraying applications are merely insurance sprays and in many cases actually unnecessary. Unfortunately, the farmer is being forced to this procedure in part by those demanding from agriculture completely insect-free produce. This has now reached ridiculous proportions. Influenced by advertising, the housewife now demands perfect specimens with no thought of or regard for how much environmental contamination has resulted to attain such perfection. If we could relax our standards to a moderate degree, pesticide contamination could be greatly reduced. Although it may be difficult to question that spraying increases yields and quality of the marketable products, there are few valid data available on how much spraying is actually necessary, how much it is adding to consumer costs, what further pests are aggravated by spraying, and what degree of resistance eventually develops.

Negative Effects on Nontarget Organisms

Although yields may be increased with greater margins of profit, according to available data, one must recognize that these chemicals may adversely affect a whole spectrum of nontarget organisms not only where applied but possibly thousands of miles from the site of application. To the ecologist concerned with the total environment, these persistent pesticides pose some serious threats to our many natural ecosystems. Certain of these are pertinent to review.

1. Killing of nontarget organisms. In practically every spray operation, thousands of nontarget insects are killed, many of which may be predators on the very organisms one is attempting to control. But such losses extend far beyond the beneficial insects. In Florida, an estimated 1,117,000 fishes of at least 30 species (20 to 30 tons), were killed with dieldrin, when sand flies were really the target organism. Crustaceans were virtually exterminated — the fiddler crabs survived only in areas missed by the treatment (Harrington and Bidlingmayer, 1958).

In 1963, there was a "silent spring" in Hanover, New Hampshire. Seventy per cent of the robin population — 350 to 400 robins — was eliminated in spraying for Dutch elm disease with 1.9 lb. per acre DDT (Wurster et al., 1965). Wallace (1960) and Hickey and Hunt (1960) have reported similar instances on the Michigan State University and University of Wisconsin campuses. Last summer, at Wesleyan University, my students observed dead and trembling birds following summer applications of DDT on the elms. At the University of Wisconsin campus (61 acres), the substitution of methoxychlor has resulted in a decreased bird mortality. The robin population has jumped from three to twenty-nine pairs following the change from DDT to methoxychlor. Chemical control of this disease is often overemphasized, with too little attention directed against the sources of elm bark beetle. Sanitation is really the most important measure in any sound Dutch elm disease control program (Matthysse, 1959).

One of the classic examples involving the widespread destruction of nontarget organisms was the fire ant eradication program in our southern states. In 1957, dieldrin and heptochlor were aerially spread over two and one-half million acres. Wide elimination of vertebrate populations resulted; and recovery of some populations is still uncertain (Rudd, 1964). In the interest of science, the Georgia Academy of Science appointed an ad hoc committee to evaluate this control-eradication program (Bellinger et al., 1965). It found that reported damage to crops, wildlife, fish, and humans had not been verified, and concluded, furthermore, that the ant is not really a significant economic pest but a mere nuisance. Here was an example where the facts did not justify the federal expenditure of $2.4 million in indiscriminate sprays. Fortunately, this approach has been abandoned, and local treatments are now employed with Mirex, a compound with fewer side effects. Had only a small percentage of this spray expenditure been directed toward basic research, we might be far ahead today in control of the fire ant.

2. Accumulation in the food chain. The persistent nature of certain of these insecticides permits the chemical to be carried from one organism to another in the food chain. As this occurs, there is a gradual increase in the biocide at each higher trophic level. Many such examples have been reported in the literature. One of the most striking comes from Clear Lake, California, where a 46,000-acre warm lake, north of San Francisco, was sprayed for pesty gnats in 1949, 1954, and 1957, with DDD, a chemical presumably less toxic than DDT. Analyses of the plankton revealed 250 times more of the chemical than originally applied, the frogs 2000 times more, sunfish 12,000, and the grebes up to an 80,000-fold increase (Cottam, 1965; Rudd, 1964). In 1954 death among the grebes was widespread. Prior to the spraying, a thousand of these birds nested on the lake. Then for 10 years no grebes hatched. Finally, in 1962, one nestling was observed, and the following year three. Clear Lake is popular for sports fishing, and the flesh of edible fish now caught reaches 7 ppm, which is above the maximum tolerance level set by the Food and Drug Administration.

In an estuarine ecosystem, a similar trend has been reported on the Long Island tidal marshes, where mosquito control spraying with DDT has been practiced for some 20 years (Woodwell et al., 1967). Here the food chain accumulation shows plankton 0.04 ppm, shrimp 0.16 ppm, minnows 1 to 2 ppm, and ring-billed gull 75.5 ppm. In general, the DDT concentrations in carnivorous birds were 10 to 100 times those in the fish they fed upon. Birds near the top of the food chain have DDT residues about a million times greater than concentration in the water. Pesticide levels are now so high that certain populations are being subtly eliminated by food chain accumulations reaching toxic levels.

3. Lowered reproductive potential. Considerable evidence is available to suggest a lowered reproductive potential, especially among birds, where the pesticide occurs in the eggs in sufficient quantities either to prevent hatching or to decrease vigor among the young birds hatched. Birds of prey, such as the bald eagle, osprey, hawks, and others, are in serious danger. Along the northeast Atlantic coast, ospreys normally average about 2.5 young per year. However, in Maryland and Connecticut, reproduction is far below this level. In Maryland, ospreys produce 1.1 young per year and their eggs contain 3 ppm DDT, while in Connecticut, 0.5 young ospreys hatch and their eggs contain up to 5.1 ppm DDT. These data indicate a direct correlation between the amount of DDT and the hatchability of eggs — the more DDT present in the eggs, the fewer young hatched (Ames, 1966). In Wisconsin, Keith (1964) reports 38% hatching failure in herring gulls. Early in the incubation period, gull eggs collected contained over 200 ppm DDT and its cogeners. Pheasant eggs from DDT-treated rice fields compared to those from unsprayed lands result in fewer healthy month-old chicks from eggs taken near sprayed fields. Although more conclusive data may still be needed to prove that pesticides such as DDT are the key factor, use of such compounds should be curtailed until it is proved that they are not the causal agents responsible for lowering reproductive potential.

4. Resistance to sprays. Insects have a remarkable ability to develop a resistance to insecticides. The third spray at Clear Lake was the least effective on the gnats, and here increased resistance was believed to be a factor involved. As early as 1951, resistance among agricultural insects appeared. Some of these include the codling moth on apples, and certain cotton, cabbage, and potato insects. Over 100 important insect pests now show a definite resistance to chemicals (Carson, 1962).

5. Synergistic effects. The interaction of two compounds may result in a third much more toxic than either one alone. For example, Malathion is relatively "safe" because detoxifying enzymes in the liver greatly reduce its toxic properties. However, if some compound destroys or interrupts this enzyme system, as certain organic phosphates may do, the toxicity of the new combination may be increased greatly. Pesticides represent one of many pollutants we are presently adding to our environment. These subtle synergistic effects have opened a whole new field of investigation. Here students of environmental science will find many challenging problems for future research.

6. Chemical migration. After two decades of intensive use, pesticides are now found throughout the world, even in places far from any actual spraying. Penguins and crab-eating seals in the Antarctic are contaminated, and fish far off the coasts of four continents now contain insecticides ranging from 1 to 300 ppm in their fatty tissues (Anonymous, 1966).

The major rivers of our nation are contaminated by DDT, endrin, and dieldrin, mostly in the parts per trillion range. Surveys since 1957 reveal that dieldrin has been the main pesticide present since 1958. Endrin reached its maximum, especially in the lower Mississippi River, in the fall of 1963 when an extensive fish kill occurred and has since that time decreased. DDT and its cogeners, consistently present since 1958, have been increasing slightly (Breidenbach et al., 1967).

7. Accumulation in the ecosystem. Since chlorinated hydrocarbons like DDT are not readily broken down by biological agents such as bacteria, they may not only be present but also accumulate within a given ecosystem. On Long Island, up to 32 lb. of DDT have been reported in the marsh mud, with an average of 13 lb. presumed to be correlated with the 20 years of mosquito control spraying (Woodwell et al., 1967). Present in these quantities, burrowing marine organisms and the detritis feeders can keep the residues in continuous circulation in the ecosystem. Many marine forms are extremely sensitive to minute amounts of insecticides. Fifty per cent of a shrimp population was killed with endrin 0.6 parts per billion (ppb). Even 1 ppb will kill blue crabs within a week. Oysters, typical filter feeders, have been reported to accumulate up to 70,000 ppm. (Loosanoff, 1965). In Green Bay along Lake Michigan, Hickey and Keith (1964) report up to 0.005 ppm wet weight of DDT, DDE, and DDD in the lake sediments. Here the accumulation has presumably been from leaching or run-off from surrounding agricultural lands in Door County, where it is reported that 70,000 pounds of DDT are used annually. Biological concentration in Green Bay is also occurring in food chain organisms, as reported at Clear Lake, California. Accumulation of biocides, especially in the food chain, and their availability for recycling pose a most serious ecological problem.

8. Delayed response. Because of the persistent nature and tendency of certain insecticides to accumulate at toxic levels in the food chain, there is often a delayed response in certain ecosystems subjected either directly or indirectly to pesticide treatment. This was the case at Clear Lake, where the mortality of nontarget organisms occurred several years after the last application. This is a particularly disturbing aspect, since man is often the consumer of those food chain organisms accumulating pesticide residues. In the general population, human tissues contain about 12 ppm DDT-derived materials. Those with meatless diets, and the Eskimos, store less; however, agricultural applicators and formulators of pesticides may store up to 600 ppm DDT or 1000 ppm DDT-derived components. Recent studies indicate that dieldrin and lindane are also stored in humans without occupational exposure (Durham, 1965). The possibility of synergistic effects involving DDT, dieldrin, lindane, and other pollutants to which man is being exposed may result in unpredictable hazards. In fact, it is now believed that pesticides may pose a genetic hazard. At the recent conference of the New York Academy of Science, Dr. Onsy G.

Fahmy warned that certain chlorinated hydrocarbons, organophosphates and carbamates were capable of disrupting the DNA molecule. It was further noted that such mutations may not appear until as many as 40 generations later. Another scientist, Dr. M. Jacqueline Verrett, pointed out that certain fungicides (folpet and captan) thought to be nontoxic have chemical structures similar to thalidomide.

We are obviously dealing with many biological unknowns in our widespread use of presumably "safe" insecticides. We have no assurance that 12 ppm DDT in our human tissue, now above the permissible in marketable products for consumption, may not be resulting in deleterious effects in future generations. As Rudd warns (1964): ". . . it would be somewhat more than embarrassing for our 'experts' to learn that significant effects do occur in the long term. One hundred and eight million human guinea pigs would have paid a high price for their trust."

Of unpredicted delayed responses, we have an example in radiation contamination. In the Bravo tests on Bikini in 1954, the natives on Rongelap Atoll were exposed to radiation assumed to be safe. Now more than a decade later, tumors of the thyroid gland have been discovered in the children exposed to these presumably safe doses (Woodwell et al., 1966). Pesticides per se or synergisms resulting from their interaction could well plague man in now unforeseen or unpredictable ways in the future.

The Sound Use of Herbicides

In contrast to insecticides, herbicides are chemical weed-killers used to control or kill unwanted plants. Following World War II, the chlorinated herbicide 2, 4-D began to be used widely on broadleaf weeds. Later, 2, 4, 5-T was added, which proved especially effective on woody species. Today, over 40 weed-killers are available. Although used extensively in agriculture, considerable quantities are used also in aquatic weed control and in forestry, wildlife, and right-of-way vegetation management. Currently, large quantities are being used as defoliators in Vietnam.

Fig. 1. Pronounced leaf curling of white oak leaves resulting from indiscriminate stem-foliar roadside spraying (left) compared to normal white oak foliage showing some insect damage (right). (Photo 1957)

Although herbicides in general are much safer than insecticides in regard to killing nontarget organisms and in their residual effects, considerable caution must be exercised in their proper use. One of the greatest dangers in right-of-way vegetation management is their indiscriminate use, which results in habitat destruction. Drift of spray particles and volatility may also cause adverse effects on nontarget organisms, especially following indiscriminate applications. In the Connecticut Arboretum, shade trees have been seriously affected as a result of indiscriminate roadside sprays (Niering, 1959). During the spring of 1957, the town sprayed the marginal trees and shrubs along a roadside running through the Arboretum with 2, 4-D and 2, 4, 5-T (1 part chemical: 100 parts water). White oaks overarching the road up to 2 feet in diameter were most seriously affected. Most of the leaves turned brown. Foliage of scarlet and black oaks of similar size exhibited pronounced leaf curling (Fig. 1). Trees were affected up to 300 feet back from the point of application within the natural area of the Arboretum. White oak twigs near the sprayed belt also developed a striking weeping habit (Fig. 2) as twig elongation occurred — a growth abnormality still conspicuous after 10 years.

The effectiveness of the spray operation in controlling undesirable woody growth indicated a high survival of unwanted tree sprouts. Black birch and certain desirable shrubs were particularly sensitive. Shrubs affected were highly ornamental forms often planted in roadside beautification programs. The resulting ineffectiveness of the spray operation was indicated by the need for

Fig. 2. White oak twigs showing distinctive weeping growth form six months after indiscriminate stem-foliar roadside spraying in spring 1957. (Photo Oct. 1957)

cutting undesirable growth along the roadside the following year.

In the agricultural use of herbicides, drift effects have been reported over much greater distances. In California, drift from aerial sprays has been reported up to 30 miles from the point of application (Freed, 1965).

Although toxicity of herbicides to nontarget organisms is not generally a problem, it has been reported in aquatic environments. For example, the dimethylamine salt of 2, 4-D is relatively safe for bluegill at 150 ppm, but the butyl, ethyl, and isopropyl esters are toxic to fish at around 1 ppm (R. E. Johnson, personal communication). Studies of 16 aquatic herbicides on *Daphnia magna,* a microcrustacean, revealed that 2, 4-D (specific derivative not given) seemed completely innocuous but that several others (Dichlone, a quinone; Molinate, a thiolcarbamate; Propanil, an anilide; sodium arsenite and Dichlopenil, a nitrile) could present a real hazard to this lower food chain organism (Crosby and Tucker, 1966).

Fig. 4. Connecticut Arboretum Right-of-Way Demonstration Area, showing highly desirable vegetation pattern created by selective use of herbicides. Directly under the wires, tall growing trees have been removed and existing herbaceous cover and low growing shrubs have been preserved. Large huckleberry clone evident in the foreground, with taller shrubs of high bush blueberry beyond. Along the edge of the lines to the forest edge, low-growing trees can also be preserved, thereby creating a valley-shaped vegetation pattern across the right-of-way. (Photo 1965)

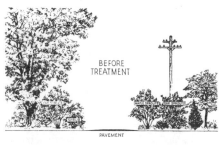

Fig. 3. Diagrammatic cross section of roadside vegetation before and after treatment. The undesirable trees, such as oak and maple, that would eventually interfere with sight line conditions along the roadside and also grow into the utility lines are cut and the stumps sprayed with the herbicide 2,4,5-T. Attractive roadside wildflowers, native shrubs, and small growing trees are preserved. (Taken from Conn. Arboretum Bull. No. 11)

Effects on rights-of-way. The rights-of-way across our nation comprise an estimated 70,000,000 acres of land, much of which is now subjected to herbicide treatment (Niering, 1967). During the past few decades, indiscriminate foliar applications have been widespread in the control of undesirable vegetation, erroneously referred to as brush (Goodwin and Niering, 1962). Indiscriminate applications often fail to root-kill undesirable species, therefore necessitating repeated retreatment, which results in the destruction of many desirable forms. Indiscriminate sprays are also used for the control of certain broadleaf weeds along roadsides. In New Jersey, 19 treatments were applied during a period of 6 years in an attempt to control ragweed (Dill, 1963). This, of course, was ecologically unsound, when one considers that ragweed is an annual plant typical of bare soil and that repeated sprayings also eliminate the competing broadleaved perennial species that, under natural successional conditions, could tend to occupy the site and naturally eliminate the ragweed. Broadcast or indiscriminate spraying can also result in destruction of valuable wildlife habitat in addition to the needless destruction of our native flora — wildflowers and shrubs of high landscape value.

Nonselective spraying, especially along roadsides, also tends to produce a monotonous grassy cover free of colorful wildflowers and interesting shrubs. It is economically and aesthetically unsound to remove these valuable species naturally occurring on such sites. Where they do not occur, highway beautification programs plant many of these same shrubs and low-growing trees.

Recognizing this nation-wide problem in the improper use of herbicides, the Connecticut Arboretum established, over a decade ago, several right-of-way demonstration areas to serve as models in the sound use of herbicides (Niering, 1955; 1957; 1961). Along two utility rights-of-way and a roadside crossing the Arboretum, the vegetation has been managed following sound ecological principles, as shown in Figures 3 and 4 (Egler, 1954; Goodwin and Niering, 1959; Niering, 1958). Basic techniques include basal and stump treatments. The former involves soaking the base of the stem (root collar) and continuing up the stem for 12 inches; the stump technique involves soaking the stump immediately after cutting. Effective formulations include 2, 4, 5-T in a fuel oil carrier (1 part chemical: 20 parts oil). Locally, stem-foliage sprays may be necessary, but the previous two techniques form the basic approach in the selective use of weed-killers. They result in good root-kill and simultaneously preserve valuable wildlife habitat and aesthetically attractive native species, all at a minimum of cost to the agency involved when figured on a long-range basis. In addition to these gains, the presence of good shrub cover tends to impede tree invasion and to reduce future maintenance costs (Pound and Egler, 1953; Niering and Elger, 1955).

Another intriguing use of herbicides is in naturalistic landscaping. Dr. Frank Egler conceived this concept of creating picturesque natural settings in shrubby fields by selectively eliminating the less attractive species and accentuating the ornamental forms (Kenfield, 1966). At the Connecticut Arboretum we have landscaped several such areas (Niering and Goodwin, 1963). The one shown in Figure 5 was designed and created by students. This approach has unlimited application in arresting vegetation development and preserving landscapes that might disappear under normal successional or vegetational development processes.

Future Outlook

Innumerable critical moves have recently occurred that may alter the continued deterioration of our environment. Secretary Udall has banned the use of DDT, chlordane, dieldrin, and endrin on Department of the Interior lands. The use of DDT has been banned on state lands in New Hampshire and lake trout watersheds in New York State; in Connecticut, commercial applications are limited to dormant sprays. On Long Island, a temporary court injunction has been granted against the Suffolk County Mosquito Control Board's use of DDT in spraying tidal marshes. The Forest Service has terminated the use of DDT, and in the spring of 1966 the United States Department of Agriculture banned the use of endrin and dieldrin. Currently, the Forest Service has engaged a top-level research team in the Pacific Southwest to find chemicals highly selective to individual forest insect pests and that

Fig. 5. A former shrub-covered thicket in the Connecticut Arboretum naturalistically landscaped with herbicides. Selective removal of less attractive species and preservation of those with highest landscape qualities have resulted in an attractive opening of red top grass surrounded by golden rods, sumac, huckleberry, chokeberry, winterberry, red cedar and other ornamental native species. (Photo 1965)

will break down quickly into harmless components. The Ribicoff hearing, which has placed Congressional focus on the problem of environmental pollution and Gaylord Nelson's bill to ban the sale of DDT in the United States are all enlightened endeavors at the national level.

The United States Forest Service has a selective program for herbicides in the National Forests. The Wisconsin Natural Resources Committee has instituted a selective roadside right-of-way maintenance program for the State. In Connecticut, a selective approach is in practice in most roadside and utility spraying.

Although we have considerable knowledge of the effects of biocides on the total environment, we must continue the emphasis on the holistic approach in studying the problem and interpreting the data. Continued observations of those occupationally exposed and of residents living near pesticide areas should reveal invaluable toxicological data. The study of migrant workers, of whom hundreds have been reported killed by pesticides, needs exacting investigation.

The development of more biological controls as well as chemical formulations that are specific to the target organism with a minimum of side effects needs continuous financial support by state and federal agencies and industry. Graduate opportunities are unlimited in this field.

As we look to the future, one of our major problems is the communication of sound ecological knowledge already available rather than pseudoscientific knowledge to increase the assets of special interest groups (Egler, 1964; 1965; 1966). The fire ant fiasco may be cited as a case in point. And as Egler (1966) has pointed out in his fourth most recent review of the pesticide problem: ". . . 95% of the problem is not in scientific knowledge of pesticides but in scientific knowledge of human behavior. . . . There are power plays . . . the eminent experts who deal with parts not ecological wholes."

One might ask, is it really good business to reduce the use of pesticides? Will biological control make as much money? Here the problem integrates political science, economics, sociology, and psychology. Anyone seriously interested in promoting the sound use of biocides must be fully cognizant of these counter forces in our society. They need serious study, analysis, and forthright reporting in the public interest. With all we know about the deleterious effects of biocides on our environment, the problem really challenging man is to get this scientific knowledge translated into action through the sociopolitical pathways available to us in a free society. If we fail to communicate a rational approach, we may find that technology has become an invisible monster as Egler has succinctly stated (1966).

Pesticides are the greatest single tool for simplifying the habitat ever conceived by the simple mind of man, who may yet prove too simple to grasp the fact that he is but a blind strand of an ecosystem web, dependent not upon himself, but upon the total web, which nevertheless he has the power to destroy.

Here environmental science can involve the social scientist in communicating sound science to society and involve the political scientist in seeing that sound scientific knowledge is translated into reality. Our survival on this planet may well depend on how well we can make this translation.

References

Ames, P. L. 1966. DDT residues in the eggs of the osprey in the northeastern United States and their relation to nesting success. *J. Applied Ecol.*, **3** (suppl.): 87-97.

Anonymous. 1966. Fish, wildlife and pesticides. U.S. Dept. of Interior, Supt. of Doc. 12 p.

Bellinger, F., R. E. Dyer, R. King, and R. B. Platt. 1965. A review of the problem of the imported fire ant. *Bull. Georgia Acad. Sci.*, Vol. 23, No. 1.

Breidenbach, A. W., C. G. Gunnerson, F. K. Kawahara, J. J. Lichtenberg, and R. S. Green. 1967. Chlorinated hydrocarbon pesticides in major basins, 1957-1965. *Public Health Rept.* **82**: 139-156.

Carson, Rachel. 1962. *Silent Spring*. Houghton Mifflin, Boston, 368 p.

Cottam, C. 1965. The ecologists' role in problems of pesticide pollution. *BioScience*, **15**: 457-463.

Crosby, D. G., and R. K. Tucker. 1966. Toxicity of aquatic herbicides to *Daphnia magna*. *Science*, **154**: 289-290.

Dill, N. H. 1962-63. Vegetation management. *New Jersey Nature News*, **17**: 123-130; **18**: 151-157.

Durham, W. F. 1965. Effects of pesticides on man. In C. O. Chichester, ed., *Research in Pesticides*. Academic Press, Inc., New York.

Egler, F. E. 1954. Vegetation management for rights-of-way and roadsides. *Smithsonian Inst. Rept. for 1953*: 299-322.

———. 1964a. Pesticides in our ecosystem. *Am. Scientist*, **52**: 110-136.

———. 1964b. Pesticides in our ecosystem: communication II: *BioScience*, **14**: 29-36.

———. 1965. Pesticides in our ecosystem: communication III. *Assoc. Southeastern Biologist Bull.*, **12**: 9-91.

———. 1966. Pointed perspectives. Pesticides in our ecosystem. *Ecology*, **47**: 1077-1084.

Freed, V. H. 1965. Chemicals and the control of plants. In C. O. Chichester, ed., *Research in Pesticides*. Academic Press, Inc., New York

Goodwin, R. H., and W. A. Niering. 1959. The management of roadside vegetation by selective herbicide techniques. *Conn. Arboretum Bull.*, **11**: 4-10.

———. 1962. What is happening along Connecticut's roadsides. *Conn. Arboretum Bull.*, **13**: 13-24.

Harrington, R. W., Jr., and W. L. Bidlingmayer. 1958. Effects of dieldrin on fishes and invertebrates of a salt marsh. *J. Wildlife Management*, **22**: 76-82.

Hickey, J. J., and L. Barrie Hunt. 1960. Initial songbird mortality following a Dutch elm disease control program. *J. Wildlife Management*, **24**: 259-265.

Hickey, J. J., and J. A. Keith. 1964. Pesticides in the Lake Michigan ecosystem. In The Effects of Pesticides on Fish and Wildlife. U.S. Dept. Interior Fish and Wildlife Service.

Keith, J. A. 1964. Reproductive success in a DDT-contaminated population of herring gulls, p. 11-12. In The Effects of Pesticides on Fish and Wildlife. U.S. Dept. Interior Fish and Wildlife Service.

Kenfield, W. G. 1966. *The Wild Gardner in the Wild Landscape*. Hafner, New York. 232 p.

Loosanoff, V. L. 1965. Pesticides in sea water. In C. O. Chichester, ed., *Research in Pesticides*. Academic Press, Inc., New York.

Matthysse, J. G. 1959. An evaluation of mist blowing and sanitation in Dutch elm disease control programs. N.Y. State Coll. of Agric. Cornell Misc. Bull. 30, 16 p.

Niering, W. A. 1955. Herbicide research at the Connecticut Arboretum. *Proc. Northeastern Weed Control Conf.*, **9**: 459-462.

———. 1957. Connecticut Arboretum right-of-way demonstration area progress report. *Proc. Northeastern Weed Control Conf.*, **11**: 203-208.

———. 1958. Principles of sound right-of-way vegetation management. *Econ. Bot.*, **12**: 140-144.

———. 1959. A potential danger of broadcast sprays. *Conn. Arboretum Bull.*, **11**: 11-13.

———. 1961. The Connecticut Arboretum right-of-way demonstration area—its role in commercial application. *Proc. Northeastern Weed Control.*, **15**: 424-433.

———. 1967. Connecticut rights-of-way — their conservation values. *Conn. Woodlands*, **32**: 6-9.

Niering, W. A., and F. E. Egler. 1955. A shrub community of *Viburnum lentago*, stable for twenty-five years. *Ecology*, **36**: 356-360.

Niering, W. A., and R. H. Goodwin. 1963. Creating new landscapes with herbicides. *Conn. Arboretum Bull.*, **14**: 30.

Pound, C. E., and F. E. Egler. 1963. Brush control in southeastern New York: fifteen years of stable treeless communities. *Ecology*, **34**: 63-73.

Rudd, R. L. 1964. *Pesticides and the Living Landscape*. University of Wisconsin Press, 320 p.

Wallace, G. J. 1960. Another year of robin losses on a university campus. *Audubon Mag.*, **62**: 66-69.

Woodwell, G. M., W. M. Malcolm, and R. H. Whittaker, 1966. A-bombs, bug bombs & us. Brookhaven National Lab. 9842.

Woodwell, G. M., C. F. Wurster, Jr., & P. A. Isaacson. 1967. DDT residues in an east coast estuary: a case of biological concentration of a persistent insecticide. *Science*, **156**: 821-824.

Wurster, Doris H., C. F. Wurster, Jr., & W. N. Strickland. 1965. Bird mortality following DDT spray for Dutch elm disease. *Ecology*, **46**: 488-499.

The Sea-Level Panama Canal: Potential Biological Catastrophe

John C. Briggs

The Isthmus of Panama comprises a major zoogeographic barrier for tropical marine animals that has stood for about three million years. The great majority of the species on either side of the Isthmus are distinct, at the species level, from those of the opposite side. The habitats on each side of the Isthmus are probably ecologically saturated so that maximum species diversity has been achieved.

A sea-level canal would provide ample opportunity for marine animals to migrate in either direction. This would probably result in the Eastern Pacific being invaded by over 6000 species and the Western Atlantic being invaded by over 4000 species. For the tropical Eastern Pacific, it is predicted that its fauna would be temporarily enriched but that the resulting competition would soon bring about a widespread extinction among the native species. The elimination of species would continue until the total number in the area returned to about its original level. (BioScience 19, no. 1, p. 44-47)

While the possibility of a sea-level canal somewhere in the vicinity of the Isthmus of Panama has been discussed for many years, its feasibility as an engineering project has become enhanced as the result of recent experimental work with nuclear devices that can be used for excavation. It appears now that the undertaking of this project will be strongly supported as soon as the current economic crisis in the United States is over. Until recently, the only facet of the plan that had drawn the attention of many biologists was the possibility of radiation damage. However, Rubinoff (1968) finally pointed out that there would be other important biological effects and gave examples of disastrous invasions that have occurred in other places as the results of human interference.

The New World Land Barrier

The New World Land Barrier, with the Isthmus of Panama forming its narrowest part, is a complete block to the movement of tropical marine species between the Western Atlantic and Eastern Pacific. This state of affairs has existed since about the latest Pliocene or earliest Pleistocene (Simpson, 1965; Patterson and Pascual, 1963) so that, at the species level, the two faunas are well separated. It has been estimated that about 1000 distinct species of shore fishes now exist on both sides of Central America but, aside from some 16 circumtropical species, only about 12 can be considered identical (Briggs, 1967).

This land barrier is also effective for marine invertebrates. Haig (1956, 1960) studied the crab family Porcellanidae in both the Western Atlantic and Eastern Pacific and found that only about 7% of the species were common to the two areas; de Laubenfels (1936) found a similar distribution in about 11% of the sponges he studied; and Ekman (1953), about 2.5% for the echinoderms. It seems, therefore, that only a very small proportion of the species in the major groups of marine animals are found on both sides of the Isthmus of Panama. The present Panama Canal has not notably altered this relationship since, for most of its length, it is a freshwater passage forming an effective barrier for all but a few euryhaline species.

With regard to the tropical waters on each side of the isthmus, there is no reason to suspect that each area is not supporting its optimum number of species. Studies of terrestrial biotas have indicated that most continental habitats are ecologically saturated (Elton, 1958; Pianka, 1966) and that islands demonstrate an orderly relationship between the area and species diversity (MacArthur and Wilson, 1967). Assuming the niches of the two marine areas are filled, achieving maximum species diversity, invasion by additional species could alter the faunal composition but should not permanently increase the number of species.

Regional Relationship

The tropical shelf fauna of the world may be divided into four, distinct zoogeographic regions: the Indo-West Pacific, the Eastern Pacific, the Western Atlantic, and the Eastern Atlantic. While the Indo-West Pacific undoubtedly serves as the primary

---

The author is Professor and Chairman of the Department of Zoology, University of South Florida, Tampa, Florida 33620. This research was supported by National Science Foundation Grant GB-4330. Helpful suggestions were received from J. L. Simon, H. H. DeWitt, and T. L. Hopkins.

evolutionary and distributional center (Briggs, 1966), the Western Atlantic Region may be said to rank second in importance. Its geographic area is larger (Fig. 1), its habitat diversity greater, and its fauna considerably richer than for each of the remaining two regions. Since the Western Atlantic species are the products of a richer and therefore more stable ecosystem, we may expect that they would prove to be competitively superior to those species that are endemic to the Eastern Pacific or Eastern Atlantic.

An examination of the faunal relationships between the Western Atlantic and the Eastern Atlantic does provide good circumstantial evidence that species from the former are competitively dominant. An impressive number have managed to traverse the open waters of the central Atlantic (The Mid-Atlantic Barrier) and to establish themselves on the eastern side. For example, in the shore fishes there are about 118 trans-Atlantic species but only about 24 of them have apparently come from the Indo-West Pacific via the Cape of Good Hope. The rest have probably evolved in the Western Atlantic and have successfully performed an eastward colonization journey across the ocean. None of the trans-Atlantic species belong to genera that are typically Eastern Atlantic. Recent works on West African invertebrate groups tend to show that an appreciable percentage of the species is trans-Atlantic (Briggs, 1967). It seems likely that the great majority of these species also represents successful migrations from the Western Atlantic.

Effect of the Suez Canal

The Suez Canal is a sea-level passage that has been open since 1869, but its biological effects are not entirely comparable to those that would occur as the result of a sea-level Panama Canal for two reasons: first, the Suez Canal connects two areas that are separated by a temperature barrier, the Red Sea being tropical while the Mediterranean is warm-temperate; second, the Bitter Lakes which form part of the Suez passageway have a high salinity (about 45 0/00) which prevents migration by many species.

Fig. 1. Tropical shelf waters of the Eastern Pacific, Western Atlantic, and Eastern Atlantic *(from Goode Base Map Series courtesy of the University of Chicago).*

Despite the above difficulties, the limited migratory movements that have taken place through the Suez Canal do provide some significant information. At least 24 species of Red Sea fishes have invaded the Mediterranean (Ben-Tuvia, 1966), 16 species of decapod crustaceans (Holthuis and Gottlieb, 1958), and several members of other groups such as the tunicates (Pérès, 1958), mollusks (Engel and van Eeken, 1962), and stomatopod crustaceans (Ingle, 1963). So there is ample evidence of intrusions into the eastern Mediterranean, but there are no reliable data that indicate any successful reciprocal migration. Furthermore, there are some indications that the invaders from the Red Sea (a part of the vast Indo-West Pacific Region) are replacing rather than coexisting with certain native species. George (1966) observed that, along the Lebanese coast, the immigrant fishes *Sphyraena chrysotaenia, Upeneus moluccensis,* and *Siganus rivulatus* may be replacing, respectively, the endemic *Sphyraena sphyraena, Mullus barbatus,* and *Sarpa salpa.*

An Ancient Event

It is now well established that in the past one or more seaways extended across Central America or northern South America for a considerable period of time, probably throughout the greater part of the Tertiary. While these oceanic connections assured the initial development of an essentially common marine fauna in the New World tropics, they operated as an important barrier for terrestrial animals. Later, perhaps about three million years ago, tectonic forces gradually produced an uplift that re-established the land connection between the two continents.

The effects of the new intercontinental connection must have been rapid and dramatic. The fossil record of this event is fragmentary but considerably better for the mammals than for the other terrestrial groups. Simpson (1965) presented an interesting and well-documented history of the Latin American mammal fauna. His findings relevant to the re-establishment of the Isthmus may be summarized as follows: (a) the full surge of intermigration took place in Pleistocene times with representatives of 15 families of North American mammals spreading into South America and seven families spreading in the reverse direction; (b) the immediate effect was to produce in both continents, but particularly in South America, a greatly enriched fauna; (c) the main migrants to the south were deer, camels, peccaries, tapirs, horses, mastodons, cats, weasels, racoons, bears, dogs, mice, squirrels, rabbits, and shrews; (d) in South America, the effect was catastrophic and resulted in the extinction of the unique notoungulates, litopterns, and marsupial carnivores; the native rodents and edentates were greatly reduced; and (e) now, South America has returned to about the same basic richness of fauna as before the invasion.

Comparatively, the invasion of Central and North America by South American mammals was not nearly so successful. The three migrants that have managed to survive north of Mexico — an opossum, an armadillo, and a porcupine — apparently occupy unique niches. Simpson (1965) noted that when ecological vicars met, one or the other generally become extinct. The dominant species that invaded South America were the evolutionary products of the "World Continent" including both North America and the Old World (the Siberian Land Bridge was frequently available).

Cutting the Isthmus Barrier

How effectively would a sea-level ship canal breach the New World Land Barrier? The engineering problems have been worked out using scale models. Although the mean sea-level is 0.77 feet higher on the Pacific side, it would have little effect compared to the effect of the difference in tidal amplitude. The tidal range on the Pacific side is often as great as 20 feet while it is usually less than a foot on the opposite side. For an open canal, it has been calculated that the tidal currents would attain a velocity of up to 4.5 knots and would change direction every 6 hours (Meyers and Schultz, 1949). Tide locks would probably be employed to regulate the currents but it seems apparent that the vast amount of fluctuation and mixing would provide ample opportunity for most of the marine animals (as adults or as young stages) to migrate in either direction.

Number of Affected Species

Data on the number of marine invertebrate species that inhabit the major parts of the New World tropics are not available. The total fauna is so rich and so many groups are so poorly known that it almost defies analysis. Voss and Voss (1955) reported 133 species of macro-invertebrates from the shallow waters of Soldier's Key, a little island (100 by 200 yards) in Biscayne Bay, Florida. The tiny metazoans comprising the meiofauna of the sediments were not sampled. Work in other areas has shown that the numbers of individuals per square meter in the meiofauna are about 100 times that of the macrofauna (Sanders, 1960). Although a complete tally of species has apparently never been made, there are indications from partial identifications (Wieser, 1960) that the number of species in the meiofauna is at least four or five times greater. For Soldier's Key, if we assume that the meiofauna is only four times richer in species, we would have a total of 665 benthic invertebrates.

Ichthyologists who have collected among the Florida Keys would probably agree that the shallow waters of Soldier's Key could be expected to yield close to 50 species of fishes. This provides an admittedly rough but useful ratio of 1:13 between the numbers of fish and invertebrate species for a small tropical locality. Although the fish fauna of the western Caribbean is not yet well known, the number of shore species can be approximated at about 600; this is probably a low estimate since we know that more than 600 exist in Florida waters (Briggs, 1958). Using the 1:13 ratio, the number of marine invertebrate species for the western Caribbean can be estimated at about 7800. Adding the fish species gives a total of about 8400 marine animal species.

The tropical Eastern Pacific possesses a less diversified fauna than the Western Atlantic. The Gulf of Panama and its adjacent waters is probably inhabited by a shore fish fauna of some 400 species. Using the 1:13 ratio gives an estimate of about 5200 species for the invertebrates and a total of about 5600 marine animal species. The great majority of tropical, shallow-water animals are very prolific and possess highly effective means of dispersal. It has been estimated that 80-85% of all tropical, benthic invertebrate species possess planktotrophic pelagic larvae (Thorson, 1966). Since the fishes are relatively mobile, it seems apparent that the great majority of the animal species under discussion would be capable of eventually migrating through a saltwater canal.

Assuming that 80% of the species on each side of the isthmus would succeed in moving through the canal, 6720 species would migrate westward and 4480 eastward. However, since we are dealing with only rough approximations, it would be more appropriate to simply estimate that we would probably witness the invasion of the Eastern Pacific by more than 6000 species and the invasion of the Western Atlantic by more than 4000 species.

Prediction

A logical prediction can be made most easily if the pertinent information given above is summarized as follows:

1) The great majority of the species on either side of the Isthmus are distinct, at the species level, from those of the opposite side.

2) The habitats on each side of the Isthmus are probably ecologically saturated so that maximum species diversity has been achieved.

3) The Western Atlantic Region includes a much larger area, exhibits more habitat diversity, and possesses a richer fauna than the Eastern Pacific or Eastern Atlantic Regions.

4) Western Atlantic species are apparently competitively dominant to those of the Eastern Atlantic — a smaller region but comparable in size and habitat diversity to the Eastern Pacific.

5) At least some of the dominant species that have invaded the Mediterranean via the Suez Canal seem to be replacing the native species.

6) When the land bridge to South America was re-established, the invasion of North American mammals enriched the total fauna. However, this effect was temporary since so many native South American mammals became extinct that the number of species soon returned to about its original level.

7) A sea-level canal would provide ample opportunity for marine animals to migrate in either direction. This would probably result in the Eastern Pacific being invaded by over 6000 species and the Western Atlantic being invaded by over 4000 species.

For the tropical Eastern Pacific, it is predicted that its fauna would be temporarily enriched but that the resulting competition would soon bring about a widespread extinction among the native species. The elimination of species would continue until the total number in the area returned to about its original level. *The fact that a large scale extinction would take place seems inescapable.* It would be difficult, and perhaps irrelevant, to attempt a close estimate of the number of Eastern Pacific species that would be lost. The irrevocable extinction of as few as 1000 species is about as appalling as the prospect of losing 5000 or more.

There is little doubt that the tropical Western Atlantic fauna would suffer far less. With the exception of a few species that may be ecologically distinct, the level of competition would probably be such that the invaders would not be able to establish permanent colonies. Some dominant, Indo-West Pacific species have been able to cross the East Pacific Barrier and establish themselves in the Eastern Pacific (Briggs, 1961). It is likely that a few of these forms would eventually find their way through a sea-level canal. In such cases, the equivalent Western Atlantic species would probably be eliminated.

Man has undertaken major engineering projects for most of his civilized history and the construction of such necessary facilities as canals, dams, and harbors will continue and expand as the human population grows larger. In this case, however, man would remove a major zoogeographic barrier that has stood for about three million years. The disturbance to the local environment would not be nearly as important as the migration into the Eastern Pacific of a multitude of species that would evidently be superior competitors. So, instead of having only local populations affected, the very existence of a large number of wide-ranging species is threatened. This poses a conservation problem of an entirely new order of magnitude.

Rubinoff (1968) assumed that a sea-level canal would be constructed and looked upon its advent as an opportunity to conduct the greatest biological experiment in man's history. As I have stated elsewhere (Briggs, 1968), this approach is unfortunate for it tends to divert attention from a vital conservation issue. The important question is: Should the sea-level canal project be undertaken at all? What is the value of a unique species — of thousands of unique species? Currently, many countries are expending considerable effort and funds in order to save a relatively few endangered species. The public should be aware that international negotiations now being carried on from a purely economic viewpoint are likely to have such serious biological consequences. Does our generation have a responsibility to posterity in this matter?

A biological catastrophe of this scope is bound to have international repercussions. The tropical waters of the Eastern Pacific extend from the Gulf of Guayaquil to the Gulf of California. Included are the coasts of Ecuador, Columbia, Panama, Costa Rica, Nicaragua, Honduras, El Salvador, Guatemala, and Mexico. While the prospect of such an enormous loss of unique species is something that the entire world should be aware of, these countries are the ones that will be directly affected since their shore faunas will probably be radically changed.

Alternative

Assuming that a better canal would provide economic benefits, I suggest either an improvement of the existing structure or the construction of a new overland canal that would still contain freshwater for most of its route. There seems to be no reason why we cannot have a canal that could accommodate ships of any size yet still maintain the freshwater barrier that is so important. One could conceive of other alternatives such as a sea-level canal provided with some means of killing the migrating animals — possibly by heating the water or adding lethal chemicals. However, such expedients would be both risky and distasteful.

References

Ben-Tuvia. Adam. 1966. Red Sea fishes recently found in the Mediterranean. *Copeia*, **2**: 254-275, 2 figs.

Briggs, John C. 1958. A list of Florida fishes and their distribution. *Bull. Florida State Museum, Biol. Sci.*, **2** (8): 223-318, 3 figs.

———. 1961. The East Pacific Barrier and the distribution of marine shore fishes. *Evolution*, **15** (4): 545-554, 3 figs.

———. 1966. Zoogeography and evolution. *Evolution*, **20** (3): 282-289.

———. 1967. Relationship of the tropical shelf regions. *Studies Trop. Oceanog. Miami*, **5**: 569-578.

———. 1968. Panama's sea-level canal, *Science*, **169** (3853): 511-513.

Ekman, Sven, 1953. *Zoogeography of the Sea*. Sidgwick and Jackson, London, xiv + 417 pp., 121 figs.

Elton, Charles S. 1958. *The Ecology of Invasions by Animals and Plants*. Methuen, London, pp. 1-181, 51 figs., photos.

Engel, H., and C. J. van Eeken. 1962. Red Sea Opisthobranchia from the coast of Israel and Sinai. *Sea Fisheries Res. Sta. Haifa, Israel, Bull.* **30**: 15-34, 7 figs.

George, Carl J. 1966. A two year study of the fishes of the sandy littoral of St. George Bay, Lebanon. *Abstr. 2nd Intern. Oceanog. Congr., Moscow*, p. 130.

Haig, Janet. 1956. The Galatheidea (Crustacea Anomura) of the Allan Hancock Atlantic Expedition with a review of the Porcellanidae of the western North Atlantic. *Allan Hancock Atlantic Exped.*, **8**: 1-44, 1 pl.

———. 1960. The Porcellanidae (Crustacea Anomura) of the Eastern Pacific. *Allan Hancock Pacific Exped.*, **24**: viii + 440, 12 figs., 42 pls.

Holthuis, L. B., and E. Gottlieb. 1958. An annotated list of the decapod Crustacea of the Mediterranean coast of Israel, with an appendix listing the Decapoda of the eastern Mediterranean. *Bull. Sea Fish. Res. Sta. Israel*, **18**: 1-126, 15 figs.

Ingle, R. W. 1963. Crustacea Stomatopoda from the Red Sea and the Gulf of Aden. *Sea Fisheries Res. Sta., Haifa, Israel, Bull.* **33**: 1-69, 73 figs.

de Laubenfels, Max W. 1936. A comparison of the shallow-water sponges near the Pacific end of the Panama Canal with those at the Caribbean end. *Proc. U.S. Natl. Museum,* **83** (2993): 441-446, 6 figs.

MacArthur, R. H., and E. O. Wilson. 1967. The theory of island biogeography. *Monograph in Population Biology 1.* Princeton University Press, New Jersey. xi + 203 pp., 60 figs.

Meyers, J. S., and E. A. Schultz. 1949. Tidal currents *in* Panama Canal—The sea-level project: a symposium. *Trans. Am. Soc. Civil Engr.,* **114:** 665-684, 40 figs.

Patterson, B., and R. Pascual. 1963. The extinct land mammals of South America. Program, XVI *Intern. Congr. Zool.,* pp. 138-148.

Pianka, E. R. 1966. Latitudinal gradients in species diversity: a review of concepts. *Am. Naturalist,* **100** (910): 33-46.

Pérès, J. M. 1958. Ascidies recoltées sur les côtes Mediterranéener d'Israël. *Sea Fisheries Res. Sta., Haifa, Israel, Bull.,* **19:** 143-150.

Rubinoff, Ira. 1968. Central American sea-level canal: possible biological effects, *Science,* **161** (3844): 857-861, 3 figs.

Sanders, H. L. 1960. Benthic studies in Buzzards Bay III. The structure of the soft-bottom community. *Limnol. Oceanog.,* **5** (2): 138-153, 3 figs.

Simpson, G. G. 1965. *The Geography of Evolution.* Chilton Books, Philadelphia & New York. x + 249 pp., 45 figs.

Thorson, G. 1966. Some factors influencing the recruitment and establishment of marine benthic communities. *Netherlands J. Sea Res.,* **3** (2): 267-293.

Voss, G. L., and N. A. 1955. An ecological survey of Soldier Key, Biscayne Bay, Florida. *Bull. Marine Sci. Gulf Carib.,* **5** (3): 203-229, 4 figs.

Wieser, W. 1960. Benthic studies in Buzzards Bay II. The meiofauna. *Limnol. Oceanog.,* **5** (2): 121-137, 4 figs.

Thermal Addition:
One Step from Thermal Pollution

Sharon Friedman

(BioScience *19*, no. 1, p. 60-61)

Many forms of aquatic animal and plant life are being threatened by the tons of heated water being spewed forth from electric and thermonuclear generating plants into the nation's rivers, lakes, and coastal waters.

Actually, the threat is more potential than real at present, but it could become a serious hazard as the power needs of the nation increase and more and more generating plants are built on America's waterways. As one scientist put it at a recent meeting on this problem, "Today we can still talk about thermal addition, but tomorrow we may be talking about thermal pollution."

More than 200 scientists representing 27 states and 6 nations recently attended the 2nd Annual Thermal Workshop of the International Biological Program to discuss the good and bad effects of thermal addition on freshwater and marine life. The 4-day meeting was held in Solomons, Maryland, at the Chesapeake Biological Laboratory of the Natural Resources Institute of the University of Maryland.

Joseph A. Mihursky, Head of the Department of Environmental Research at the Chesapeake Laboratory and Chairman of the IBP Workshop, explained that the problems caused by thermal addition have been on the rise and will continue to increase in the future. In 1958, the electricity demand was expected to quadruple by the year 2010. In 1965, estimates for the same time period indicated that the electricity demand would increase 32 times. Now these estimates have soared to as high as 256 times the present need by 2010. This would mean that even by 1980, electricity production may amount to 2 million megawatts. If engineering designs remain the same, this production would require 200 billion gallons of cooling water per day — the equivalent of one-sixth of the total volume of fresh-water runoff per day in the United States. During dryer seasons of the year, half of the total runoff would be needed for cooling purposes.

Even now, Dr. Mihursky said, the problem is bad in heavily populated and industrialized northeastern U.S. watersheds where, due to re-use, more than 150% of freshwater flow passes through the various steam electric stations during the summer low-flow periods.

"Power plants," he continued, "generally increase water temperatures 10 to 30 degrees Fahrenheit above normal with discharges commonly reaching 100-115 degrees during summer months. Temperatures of one northeastern river have reached 140 degrees due to heated discharges from various industries — a level of 40-50 degrees above the tolerance of most aquatic organisms."

The advent of thermonuclear generating plants will increase the thermal addition problem since these plants have large generating capacities ranging from 1000 to 4000 megawatts. Nuclear power plants are less efficient than traditional ones and produce about 50% more waste heat per unit of electricity. These plants may each require up to 7 billion gallons of water a day for cooling. Six months ago, the Atomic Energy Commission reported that 15 nuclear power plants were in operation in this country and that 87 more were either under construction or planned for the immediate future. Not included in these figures were the many proposed plants that are not far enough along to have applied for building permits.

Although the AEC keeps tight rein on radiological hazards and safeguards of nuclear plants, it has no responsibilities regarding thermal wastes. New legislation may become necessary to clarify the authority of the various governmental agencies, and Senator Edmund Muskie's Subcommittee on Air and Water Pollution began hearings on thermal pollution in February 1968. State governments have also become concerned and have begun to establish seasonal temperature regulations under the new Federal Water Quality Act.

Heated water is not the only problem for aquatic life brought on by the power plants. There are also metallurgical effects caused by loss of metals from power plant condenser tubes and biocidal problems caused by detergents, acids, and chlorine used to keep condenser tubes clean of fouling from biological growths and deposits. The nuclear plants may present another factor — radioactive materials.

Different environments and water temperatures mediate the effects of thermal addition. Wheeler North of the California Institute of Technology pointed out at a press conference following the IBP workshop that in the cooler waters of California and also those around England and Scotland, there is not much of a problem. But, he said, this does not mean we aren't concerned, since a buildup of power plants could radically change the situation.

Many species of aquatic life have different levels of temperature tolerance. In discussing the primary producers — the grasses of the sea — Robert Krauss of the University of Maryland said a few degrees of higher temperature could make a difference, both in type and quantity, in the productivity of these organisms, which provide food and fuel for the entire ecological system.

Invertebrates, which include many economically important shellfish, also react to thermal addition, according to Jack Pearce of the U.S. Department of Interior's Sandy Hook Marine Laboratory and co-chairman of the IBP workshop. But whether the reaction is beneficial or harmful depends on the latitude of their habitats and their exposure to sun during low tides among other factors.

Charles Coutant of Battelle Memorial Institute's Northwestern Laboratory said that while all fishes face some danger from thermal addition, the migratory fishes are particularly susceptible in the mixing zones where the heated water flows into cooler natural waters. Although no large fish kills can yet be claimed on thermal addition, Dr. Coutant said, there is some question as to whether certain fish could carry out their life cycle in slightly elevated temperatures, and what these higher temperatures would do to their food requirements and behavioral characteristics.

Thermal addition can be avoided by the use of cooling towers where water used in cooling is cooled, in turn, by either the atmosphere alone or with the aid of giant fans. This method is commonly used in England, pointed out R. A. Beauchamp from the Central Electricity Research Laboratory in Surrey. This is the only way it is possible to have so many generating plants located in such a small area, he said, without doing severe damage to the environment.

Cooling towers have not found favor in the United States because of their added building and maintenance expense — sometimes amounting to millions of dollars. These towers also have other drawbacks: they are huge and unsightly, and one type — an open-circuit variety — loses a significant amount of water through evaporation to the atmosphere, sometimes causing fog and ice to develop as side effects.

With more knowledge of its effects, thermal addition may be put to good use in the future. Experimenters from Scotland reported using the heated waters from a power plant to encourage production of flounder, sole, and some shellfish. Although not enough experimentation has been done to be conclusive, most scientists at the workshop agreed that thermal addition could be looked upon as a resource which, under proper conditions, could be used to spur fish production in cold waters.

The major recommendation of the IBP workshop, according to Dr. Mihursky, was that biologists, ecologists, and engineers must work together to determine the locations of power plants and their design. By this cooperation, the electric power industry can still fill the growing power needs of the nation without necessarily endangering its freshwater and marine plants and animals. Another recommendation was for the continuation of this workshop series because of the great need for scientific data on the effects of thermal addition and for greater collaboration among scientists all over the world on this problem. As Lionel A. Walford, Director of the Sandy Hook Marine Laboratory, said: "Thermal addition affects man's food supply and is therefore a world problem. It is fitting that research on it is included in the International Biological Program where scientists from 55 nations are searching for understanding of man and his environment."

DDT on Trial in Wisconsin

Bruce Ingersoll

(BioScience *19*, no. 4, p. 357-358)

"The use of DDT is wrong — dead wrong," contends University of Wisconsin ecologist Joseph J. Hickey. Just how wrong, the Environmental Defense Fund (EDF) is trying to prove in hearings on a petition to ban dichlorodiphenyltrichloroethane (DDT) in Wisconsin. The EDF, a militant, new force on the conservation front, is using the quasi-judicial hearings in Madison as a national forum for anti-DDT testimony.

"Wisconsin is the showdown," says biologist Charles F. Wurster, Jr., of the State University of New York at Stony Brook. The showdown is with the American Agricultural Chemicals Association's "task force for DDT." During 3 weeks of testimony, Wurster, EDF attorney Victor J. Yannacone, Jr., of Patchogue, New York, and their impassioned colleagues sought to overwhelm the Association's political influence and financial might with scientific facts. When the hearings resume, the chemical pesticide establishment will present its defense of DDT, the cheapest of all modern pesticides.

The Wisconsin Izaak Walton League and the Citizens Natural Resources Association set up this test case by asking the state Department of Natural Resources (DNR) to declare DDT a water pollutant and outlaw its use wherever it can contaminate state waters. Should the Wisconsin DNR do so, other states may follow suit. (So far, only the Arizona Pesticide Control Board has banned the use of DDT. However, the ban is for one year and covers only commercial agriculture. It was motivated by fears that DDT residues in Arizona produce might exceed FDA tolerances. The Arizona Board did not condemn DDT as an environmental pollutant — which is the ruling that EDF hopes for in Wisconsin.) It is significant that during the proceedings the Wisconsin Department of Agriculture decided to quit recommending DDT for Dutch elm disease control — a partial victory for the foes of DDT. Moreover, on the basis of the testimony so far, a bill has been introduced in the Wisconsin Legislature to ban DDT outright. Wisconsin's Senator Gaylord Nelson fired the opening salvo against DDT: "In one generation DDT has contaminated the atmosphere, the sea, the lakes and streams, and infiltrated the tissues of most of the world's creatures." A DDT ban, he testified, would be a "landmark step forward" and certainly the "first ray of hope" for many endangered wildlife species. Nelson was followed to the stand by a dozen top-flight scientists, rounded up from all over the nation by Wurster. Their testimony was guided by the flamboyant Yannacone, who at age 32 is rapidly making a national name for himself in his court contests with "polipollutionists."

Wurster enumerated the strikes against the chlorinated hydrocarbon DDT. Being a very stable chemical, he testified, it persists in the environment for 10 years or more. It is nearly insoluble in water and cannot be dissipated in the oceans. Since it is extremely soluble in fatty tissue, it builds up in the fat of fish, wildlife, and livestock. Moreover, it is highly mobile. DDT, he said, is a vagabond for it tends to hitch rides on particles suspended in flowing water, to attach itself to soil particles easily swept aloft by wind, and to evaporate with water into the atmosphere. And this insecticide has other idiosyncracies which enable it to pervade the biosphere. It moves up a food chain as one organism eats another, becoming more concentrated or "magnified" with each link, eventually manifesting its sublethal effects. In killing insect pests, it often kills nontarget species.

Robert van den Bosch, an entomologist from the University of California, Berkeley, said that DDT is such a "crude ecological poison" that it sometimes wipes out the natural predators that would keep insect pests in check.

In sum, Wurster said, "there cannot be controlled use of an uncontrollable compound." He fears DDT will kill off the rare Bermuda petrel, an oceanic bird believed extinct for 300 years before a few were found off Bermuda some 25 years ago. The survival rate of petrel chicks dropped from 67% in 1958 to 36% in 1967 because of DDT poisoning. "If this trend continues," Wurster warned, "reproductive success will reach zero by 1978." DDT residues in petrel eggs indicate that the Atlantic basin "contains substantial DDT contamination — cause for alarm since DDT has been found to interfere with photosynthesis by marine phytoplankton. Photosynthesizing phytoplankton produces 70% of the oxygen we breathe, he said.

"Lake Michigan is one of the most DDE-polluted lakes in the world," Hickey testified. In illustrating the extent of the pollution, he outlined how dichlorodiphenyldichloroethylene (DDE), a DDT metabolite, is biologically magnified in an aquatic ecosystem off Wisconsin's Door County peninsula. One-quarter of one part per million (ppm) of DDE was detected in tiny crustaceans low on the food ladder. Whitefish averaged 3 ppm of DDE, while fish-eating herring gulls on the top rung had 80 ppm in their muscle tissue and 1925 ppm in their fat. Thus, the ultimate victims of DDE magnification are species atop food ladders, particularly such birds of prey as the osprey, peregrine falcon, and bald eagle.

Hickey and Berkeley's Robert W. Risebrough, a molecular biologist, explained that DDE causes a bird's liver to produce enzymes that break down female sex hormones and thereby inhibit calcium metabolism. Unable to mobilize enough calcium, birds lay thin-shelled eggs which crack or break easily. Subsequent failure to reproduce has caused drastic population "crashes" among these species after 1950. The peregrine falcon, for instance, has been extirpated in the eastern half of the United States. Where 45 pairs of bald eagles used to nest on the shores of Lake Michigan in the 1940's, just one pair could be found last summer; and the pair has not reproduced since 1964.

Yannacone called Lucille F. Stickel, pesticide research coordinator at the Department of the Interior's Patuxent (Maryland) Wildlife Research Center, to the stand to corroborate Hickey's testimony that DDE is "a chemical compound of extinction." She said that in one experiment 3 ppm doses of DDE in the diets of mallard ducks were found to decrease egg shell thickness, increase egg breakage, and sharply reduce overall reproductive success. In another Patuxent study, kestrels suffered marked reproductive failure when fed mixtures of DDT and dieldrin.

DDT is not just an avian problem. Kenneth Macek, biologist with the Department of Interior's fish-pesticide research laboratory at Columbia, Missouri, testified that brook trout showed much higher mortality *after* being exposed to DDT for 6 months. He found in a New Hampshire study that 96% of the trout fed 3 mg of DDT (per kilogram of fish per week) succumbed in the next 3 months to such environmental stresses as sharp changes in water temperature and food shortages, while 99% of the control trout survived. *Direct* DDT poisoning, Macek noted, has been blamed for the death of coho salmon fry in Michigan hatcheries. DDT residues were discovered in the egg-yolk sacs on which fry rely for sustenance until they can fend for themselves. Because the eggs came from Lake Michigan salmon, the die-off does not augur well for Michigan's annual fall "coho boom" or Wisconsin's coho-stocking program.

Much of the DDT tainting Lake Michigan and its fish is aerial "fallout," Risebrough said. Air-borne particles laden with DDT settle out much like radioactive dust from nuclear tests. DDT has been found in rain and snow, in atmospheric samples taken over the Indian Ocean, and off San Francisco and the Caribbean island of Barbados.

Biochemical pharmacologist Richard M. Welch, another EDF witness, said that DDT acts much like estrogen, a sex hormone found in animals and man. A DDT injection of 5 mg causes the uterus in laboratory rats to enlarge and become heavier, he testified. It also increases the activities of enzymes in a rat liver that attack the sex hormones testosterone and estradiol. "If we can extrapolate data from animals to man, then we can say this increase in enzyme activities may also occur in man," Welch said. In questioning Welch, who works for Burroughs Wellcome Research Laboratory in Tuckahoe, New York, Yannacone emphasized that man has been the unwitting guinea pig for an uncontrolled experiment with DDT.

Louis A. McLean of Northfield, Illinois, a lawyer in the DDT camp, dismissed the testimony to date as the opinion of "young men outside the areas of their expertise." The record will be set straight, he implied, when DDT's defenders get a chance to call the "expert" witnesses. Willard Stafford, a very successful trial attorney from Madison, is expected to direct the defense. He is certain to stress DDT's relative effectiveness and its admirable record in suppressing insect-carried typhus and malaria.

Yannacone and the other EDFers are anxious for the hearings to resume. They fear that spring will arrive and more DDT will be unleashed on the landscape before the Department of Natural Resources can review the testimony and make a declaratory ruling. Whatever the outcome, a court appeal appears inevitable. EDF hopes to have the wherewithal to continue "doing the same thing the NAACP did" in civil rights — establish in the courts a body of common law whereby citizens can assert their rights to "a clean, healthy and viable environment."

Who Needs DDT?

Thomas H. Jukes

(BioScience *19*, no. 7, p. 640-641)

Many articles and letters, especially in the newspapers, have appeared in recent months regarding DDT. Most of these have favored its banning from further use because of the presence of residues in various animals, the storage of DDT in fatty tissues, its persistence in the soil, etc. At times, the opposition to the use of DDT has taken on the nature of a crusade. Thus, in *BioScience,* (April, 1969) Bruce Ingersoll states, ". . . their impassioned colleagues sought to overwhelm the (American Agricultural Chemical) Association's political influence and financial might with scientific facts."

I do not know how much "political influence and financial might" the Association has. However, those who have been representing the Association at the Wisconsin hearings tell me that they are getting very little support, and have asked me to send any information I could gather. It is my impression that the Association is not putting much effort into defending DDT.

What are the "scientific facts" which will "overwhelm" DDT? First, DDT has probably saved and extended more human lives than any man-made chemical substance in history. DDT bears the major credit, or blame for the population explosions in India and several other tropical countries. The use of DDT prevented a post-World War II typhus epidemic from taking place in Europe. In addition to its role in killing the arthropod vectors of some of the major diseases of mankind, DDT is one of the chief insecticides used in the protection of food and field crops. No authentic cases of death or serious injury to human beings from the routine use of DDT have been reported, although there have been a few industrial accidents, or freak incidents such as using DDT by mistake for pancake flour.

A major defect in much of the published material attacking DDT is the failure to consider the most elementary principle of toxicology, stated centuries ago by Paracelsus as "All things are poisonous, yet nothing is poisonous," and reiterated on 22 May 1969, in a letter to me from Professor Wayland J. Hayes, Jr. as follows:

> The basic difficulty, as I see it, is that few educated people have any idea of even the most rudimentary principles of pharmacology or toxicology. Dosage response relationships mean absolutely nothing to them. They are, therefore, not in a position to evaluate contradictory statements and are likely to mistake dedication for knowledge.

As an example ten Antarctic Adelie penguins were analyzed for DDT. Four showed traces in their body fat; average for the group, 0.05 ppm. Thirty-eight tissue samples were negative for DDT. Twelve Lake Michigan herring gulls in a "seemingly healthy condition" were killed by gunfire at their nesting colonies by members of the Department of Wildlife Management, University of Wisconsin. Average DDT in body fat was 390 ppm and total chlorinated hydrocarbons were 2441 ppm. Now if the DDT had been, for example, fluorine, the conclusion by a toxicologist would have been that, since the penguins contained, at the most, one seven-thousandth as much fluorine as "seemingly healthy herring gulls," the amount in the penguins is of no biological significance whatever. Yet the finding of this minute trace of DDT in penguins by fantastically delicate analytical procedures has become a major conservation issue; apparently because it is thought to indicate world-wide pollution, when actually the report is only an illustration of the fact that contamination can be found anywhere, provided that detection methods are sufficiently discriminatory.

Again quoting Mr. Ingersoll's article:

> DDT, he said, is a vagabond for it tends to hitch rides on particles suspended in flowing water, to attach itself to soil particles easily swept aloft by wind, and to evaporate with water into the atmosphere. And this insecticide has other idiosyncrasies which enable it to pervade the biosphere. It moves up a food chain as one organism eats another, becoming more concentrated or magnified with each link, eventually manifesting its sublethal effects.

This thesis, expressed on various planes of lyricism, has been one of the main criticisms of DDT. Dr. J. M. Barnes, who is Director of the Toxicology Research Unit of the British Medical Research Council, has voiced a contrary opinion as follows:

> Unfortunately DDT is relatively slowly metabolised and excreted by the mammal and by virtue of its solubility characteristics tends to get laid down in tissue fat. Here it would have remained as an innocent and unrecognized passenger but for the fact that the chemists invented a sensitive chemical method since further enhanced by the gas-liquid-chromatographic technique capable of detecting the chlorine and indicating its source even in minute quantities. Thus it has become possible to establish

---

EDITOR'S NOTE: This article by Dr. Jukes is published in accordance with *BioScience's* policy of presenting all sides of an issue. In August we will publish Bruce Ingersoll's second report on the DDT hearings in Wisconsin.

an anxiety neurosis in respect to a few parts per million of a compound in a tissue such as fat where a few parts per thousand in the whole animal are of no toxicological significance.

The difference between these two viewpoints needs resolution by experimental procedures, in which quantitation is rigorously applied. It is known that DDT is metabolized by bacteria, by DDT-resistant houseflies, by dairy cattle and by human beings. Storage of DDT, therefore, presumably reaches a plateau depending on intake. The effects of stored DDT on enzyme systems need evaluation in terms of net result; for example, reproduction in rats was not found to be affected by DDT in the diet at a level of 50 ppm.

A dramatic picture of an eaglet beside an unhatched egg appeared on the cover of *Science*, 7 February 1969. The failure of the second egg to hatch was attributed to an abnormally thin eggshell caused by DDT in the maternal diet. In experiments with chickens and quail hatchability has been normal even at levels of 100 or 200 ppm of diet. The quail offspring from the mothers on the 200 ppm diet had a mortality of 87% in six weeks as compared with 17% for the control group. In a controlled study of shell thickness it was found that the shells of eggs laid by Bengalese finches receiving DDT were significantly thicker than those of controls. Again, the level of DDT must be measured before generalizations are made extending to free-living wild birds. Another point is that the skeleton of the developing avian embryo receives its major calcium supply from the eggshell, which for this reason becomes markedly thinner during the incubation period. I have supported the opinion of R. L. Rudd and R. E. Gennelly (California Department of Fish and Game Bulletin, 1956) that:

> The spread of suburbs, industrial pollution, the drainage of marshlands, the building of superhighways, the increase in numbers of people, all have a disrupting effect on the wildlife population compared with which pesticides are of minor significance.

I have also proposed that spraying marshes with DDT may disturb the "balance of nature" so that certain species of birds are no longer kept in check by avian diseases such as malaria, Newcastle disease and fowlpox. The great increases in redwing blackbirds, grackles and cowbirds may have resulted from this.

The real problem regarding DDT is posed in the editorial by Sax in *BioScience* (April, 1969). It is: Should the means for preventing hunger and disease be withheld from impoverished human populations, especially in tropical countries? Here, my conditioning makes me say "no." I have stated:

> Most important of all, DDT is needed by the millions of 'Third World' people, because it is a cheap, safe residual pesticide. Professor George Nelson sent me a picture of an African with 'river blindness', caused by onchocerciasis, which, he says afflicts between 30 and 40 million in Africa. In some villages, the adults are all blind, and are led around by children who face the prospect of blindness. The disease is caused by a microscopic parasitic worm, carried by a blackfly that breeds in swiftly-flowing streams. It was accidentally found that the fly larvae in streams are readily killed by DDT, and, as a result of this discovery, the black-fly was eliminated from a large area in Kenya. This should break the cycle of infection, which depends on the parasite shuttling between people and flies. Is there hope for the children of the victims of river blindness?

Unless we are willing to forego medical treatment for ourselves and our own families, it seems to me that we must seek methods of population control other than those of lifting the curbs on disease and on destruction of crops by insects.

"OPERATION MOSQUITO IN THE HEART OF PERSIA. First victims, the children. Small boy of the Gourjii Tribe, Iran, sick with malaria."

Philip Boucas

DDT on Trial in Wisconsin - Part II

Bruce Ingersoll

(BioScience *19*, no. 8, p. 735-736)

DDT's defenders have made their stand in Wisconsin. They had their say in the second, and last, round of the closely watched DDT hearings. The "industry task force" formed by the National Agricultural Chemicals Association has long since decamped, as has the opposition, the New York-based Environmental Defense Fund (EDF). But this isn't to say the militant EDFers and well-financed pesticide people won't regroup their respective contingents of attorneys, p.r. men, and expert witnesses to fight again, conceivably in a Wisconsin court room.

The two sides are now awaiting the decision, having argued exhaustively over whether the Wisconsin Department of Natural Resources should outlaw the one-time wonder chemical as a water pollutant. The ban on DDT is being sought by the Citizens Natural Resources Association and the Wisconsin chapter of the Izaak Walton League, petitioners in the quasi-judicial hearings. For members of these groups and the conservation-minded everywhere, round one was a time when their doubts about DDT crystallized, when their worst fears were substantiated by scientific fact. Reports of testimony against DDT and its metabolites were then disseminated across the nation by journalists clambering aboard the accelerating band wagon of environmental awareness. Many people heard the indictments and tuned out, not waiting to hear the defense.

Nowadays it is less than camp to be in favor of DDT. To a certain extent, the press agent for the task force was right: DDT is the whipping boy, justly or not, of conservationists. As a target, "it's handy; it even fits into headlines easily." Politicians, sensing the nation's changing mood, call for outright bans and schedule

The first part of Mr. Ingersoll's article appeared in the April issue on p. 357.

hearings. Sen. Gaylord Nelson keeps sniping away from Washington, urging a federal ban. Anti-DDT legislation is pending in some states and the use of DDT has been barred or limited in a few others. Sweden has become the first nation to ban DDT (for 2 years). Since last March, the Food and Drug Administration has seized 38,000 pounds of DDT-tainted Lake Michigan coho salmon. As Samuel Retrosen, president of the Montrose Chemical Corp. of California, testified during round two, the tide has been running against DDT ever since Rachel Carson's *Silent Spring* appeared in 1963. DDT production was at its peak then— 183 million pounds—and by 1967 it shrank to 103 million, 70% of which was exported. Of the five chemical companies manufacturing DDT, Retrosen's stands to lose the most, since it now produces half of the DDT made in the United States.

Given the milieu of the hearings and the sentiments that seemed to prevail, DDT's defenders had an unenviable job. They were, willy-nilly, on what had become the wrong side, deny it as they vehemently would. What's more, they couldn't pursue it with an air of moral righteousness, for that the EDFers had. Even their five-man p.r. corps seemed somewhat half-hearted; the "snow-job" the EDF warned about never materialized. Did that belie a lessening of resolve in the DDT camp? A feeling that it was a lost cause? Or did it indicate a lack of anything to say in DDT's behalf? Rather, a lack of anything fresh and new to publicize, so as to counteract the ink given EDF arguments in round one?

Admittedly, some of the anti-DDT testimony had been heard before, but the bulk of it was newer to the newsmen, and thus more newsworthy than the testimony of witnesses brought in by the task force. In retrospect, the arguments for DDT were quite predictable: DDT's

drum has been beaten many times before. And, as one EDFer said, "There are only a certain number of things that can be said for it." The defense, nevertheless, was ably handled by Willard Stafford, the Madison trial lawyer retained at the end of round one. The first witness he called was Dr. Wayland J. Hayes, Jr., a well-known Vanderbilt University toxicologist and an official of the United Nation's World Health Organization.

As expected, Hayes credited the insecticide with the worldwide suppression of malaria, plague, and typhus. Malaria claimed two million lives a year in the pre-DDT era, yet that toll has now been reduced to 15,000, he testified. Were a widespread ban instituted in this country, he said, it would have a "disastrous effect on the rest of the world." He fears that underdeveloped nations would misinterpret such a development, quit using DDT, and that epidemics would ensue. This contention was treated by the EDF attorney, Victor J. Yannacone, Jr., as irrelevant to the issue at hand: does DDT pollute the waters of Wisconsin? The same goes for the accounting of DDT's admirable record in disease vector control.

Also predictable was Hayes' testimony that there is no conclusive scientific evidence that DDT imperils our health. He asserted that the coho salmon impounded by the FDA because of DDT levels as high as 19 ppm would have been safe to eat. He based this judgment on two DDT studies he supervised as chief of the U.S. Public Health Service's toxicology section. Hayes described experiments with Georgia convict volunteers who were fed 3.5 mg or 35 mg of DDT per day for 12 to 21 months in the early 1950's, and recent tests on workers at a California DDT-manufacturing plant. The prisoners given 35 mg doses—200 times the amount then ingested by the average person— had 234 to 281 ppm in their body fat, whereas fat levels for the general population ranged from 2.3 to 4 ppm. As for the factory workers, he said they were exposed to 17 to 18 mg per day, some of them for as long as 19 years, compared with the average exposure of 0.028 mg per day. Yet, in both cases, DDT was found to have no apparent effect on the health or work performance of the men, according to Hayes.

Under cross-examination by Yannacone, Hayes admitted that neither study explored DDT's effect on enzyme production by the human liver. Yannacone consequently pressed Hayes—rather inconclusively—about the possibility of DDT damage to the liver, as suggested by Dr. William Deichmann, a University of Miami pharmacologist who recently reported finding far more DDT than normal in the fat, brain, and liver tissue of 201 victims of liver disorders. Yannacone, alternately lolling in his chair or lunging up to make a point, was particularly dogged in cross-examining Hayes. The two of them went round and round for hours, arguing over the thoroughness of the two PHS studies and what Hayes exactly meant by calling DDT absolutely safe. At one point, Yannacone blew up and accused Hayes of evading his questions. Hearing examiner Maurice Van Susteren quelled and then reprimanded Yannacone for "histrionics" and "badgering the witness."

In the rhubarb over semantics, Hayes eventually defined safe as "showing no clinical signs of effects." He did concede that "DDT acts physically on the nerves, but there is not enough evidence to know what happens. Hayes disagreed with rebuttal testimony by neurophysiologist Alan B. Steinbach from the Albert Einstein College of Medicine in New York City. Steinbach said DDT is unique among poisons in that its effects on the nervous system are irreversible. Hayes countered, "Clinical effects of DDT are clearly reversible . . . When DDT is removed, changes regress."

DDT's defenders raised the possibility that DDT may be taking the rap for PCB (polychlorinated biphenyl), a plasticizing agent widely used by industry. It has been known since 1967 that PCB can be mistaken for DDT or DDD in gas chromatography tests for pesticide residues, according to Francis B. Coon, head of the Wisconsin Alumni Research Fund (WARF) Institute's chemistry department. He testified that a recheck of a test of a coho salmon indicated 70% of the DDT-DDD residue might have been PCB. No doublecheck was done on DDE, because it takes weeks and isn't necessarily conclusive, he said. Yannacone questioned him closely about WARF lab procedures and even got him to admit that he had conducted comparatively few tests with a gas chromatograph in recent years.

To challenge first round testimony that DDE has caused drastic reproductive failure among birds of prey by disrupting their calcium metabolism and making them lay thin-shelled and easily broken eggs, Stafford called Frank Cherms, a University of Wisconsin poultry geneticist, to the stand. Cherms reported he fed DDT to five generations of Japanese quail in hopes of developing a resistant strain, yet found no signs of thin egg shells nor changes in reproduction. Egg-shell thickness, can be affected by a host of factors—heredity, nutrition, environmental stress (fright and excessive heat), and disease (avian influenza, bronchitis, and Newcastle disease).

Another dimension to the pro-DDT case was added by William F. Gusey, wildlife consultant for Shell Chemical Co. To offset DDT-isscourge-of-wildlife testimony, he cited increase in mammal populations and encouraging trends in small game and upland bird populations in the last 30 years.

Rutgers University entomologist Bailey B. Pepper contended that DDT is indispensable in some cases. The former president of the Entomological Society of America testified that DDT is needed for mosquito control in his home state of New Jersey where a reservoir of mosquito-borne encephalitis persists among animals and human cases of encephalitis were reported last year. Moreover, the cotton bollworm—known as the tomato fruitworm in gardens and the corn earworm in corn fields—is extremely hard to control without resorting to DDT. Other insect control methods leave something to be desired, Pepper said. For example, use of biological controls, chiefly predaceous insects, hasn't proved too dependable against some crop pests.

The economic argument came from R. Keith Chapman, the University of Wisconsin entomologist who testified that Wisconsin, thanks to DDT, could grow vegetables "the likes of which have never been seen before." He characterized DDT as the safest, most effective, and most economical insecticide available for many vegetable crops. DDT has given "miraculous results," he said, in checking the cabbage looper and the potato leafhopper, as well as aster yellows disease in several varieties of vegetables and ornamentals. If it were outlawed, Chapman predicted Wisconsin farmers would lose 2 to 3 million dollars. Hardest hit would be the carrot growers, he said, "for they would be put at an economic disadvantage if they had to compete with other areas that did not have a ban." Carrots are a 2.3-million-dollar-crop in Wisconsin, the nation's No. 1 producer of vegetables for processing.

There was one witness to whom Stafford objected strenuously and whom Yannacone would not claim as his, although EDF was paying his expenses. He was S. Goran Lofroth, a Swedish toxicologist and head of the University of Stockholm's radiobiology department. Lofroth was ultimately called to the stand by assistant attorney general Robert McConnell, the state's public intervenor in the hearings. He testified that many breast-fed babies in the world are getting 0.02 mg of DDT per kilogram of body weight—twice the maximum daily dose recommended by the World Health Organization. This rate of intake is "in the range where laboratory animals show pharmaco-dynamical changes," he said. "What these changes mean is not known. One cannot predict the consequences if these and similar changes work in man and one does not know what the future holds for persons exposed to that much DDT." Lofroth urged a moratorium on the use of DDT at least until more research can be done on its possible perils to man.

The decision on the petition for a declaratory ruling against DDT should be forthcoming sometime this fall, if ever. (Conceivably, the Natural Resources Board might find it politically expedient to sit on the matter indefinitely. There is some question whether it has the power under state law to make such a ruling, too.) At the conclusion of the hearings, attorneys for both sides exhorted the board to let Van Susteren make the decision. Odds are considerably better, however, that Van Susteren will summarize the testimony in the 3500-page transcript for the board or perhaps submit a recommendation and that DNR secretary Les P. Voigt will act on behalf of the six other board members. As one DNR official sees it: "Wisconsin still has a chance to make a significant decision—declaring DDT a water pollutant could be a real precedent."

"Whether one side wins or loses," he said, "the public is making a decision on DDT."

DDT Roundup

(BioScience *19*, no. 8, p. 739)

In mid-July the Wisconsin state assembly unanimously passed and sent to the senate a bill banning the sale and distribution of DDT in Wisconsin. Senator *Gaylord Nelson* (D-Wis), who has introduced legislation in every Congress since the 89th to ban the use of DDT, commented, *"The action of the Wisconsin assembly is one of the final steps to victory for the scientists, conservationists, and concerned public who have contested the use of DDT in our state."* Many of these people were involved in recent public hearings before the Wisconsin Department of Natural Resources which focused national attention on the confrontation between the pesticide industry and anti-DDT forces over the use of the controversial chemical (April *BioScience*, p357 and this issue, p735). The Wisconsin action represented a victory for the anti-DDT faction who not only succeeded in foiling an attempt by DDT-proponents to table the bill but also pushed through the present version over a milder one supported by the Wisconsin Farm Bureau Federation. The measure was amended to permit a "DDT emergency board" which could authorize use of the chemical in cases where a *"significant portion"* of a crop is affected by epidemic plant disease or in cases of an epidemic disease of humans or animals. The board will be composed of the Wisconsin secretary of agriculture, the state health officer, and the secretary of natural resources. Although as we go to press the bill has not been acted upon in the senate, a spokesman for Senator Nelson indicates that there is a consensus that the legislation should not encounter too many problems in a speedy passage through the senate.

Thus, Wisconsin joins a growing number of states and countries either banning or actively reviewing the use of the pesticide. Sweden, Denmark, and Hungary have banned DDT as have the states of Arizona and Michigan and an array of bills relating to pesticide use has been introduced in various state legislatures. Last month Secretary-General *U Thant* of the United Nations transmitted a study to member nations reporting that a billion pounds of DDT had been spread through the environment and that world production of pesticides of one kind or another amounted to 1.3 billion pounds per year.

On 9 July Acting Secretary of Agriculture *J. Phil Campbell* ordered a 30-day suspension of pest control programs conducted by the Department, pending a review of these activities. The suspension order affects programs of the Agricultural Research Service and the Forest Service involving any planned application of DDT, dieldrin, endrin, aldrin, chlordane, toxaphene, lindane, heptachlor, or BHC. Possible alternative control methods, including other chemicals which can be used in place of the persistent pesticides, will be considered. As a result of the order, Department pest control operations on military and civilian airports will be suspended, along with cooperative federal-state projects, and applications of persistent pesticides to national forests. In announcing the action, Acting Secretary Campbell cited a recent report requested by the Department from the National Academy of Sciences–National Research Council, recommending further steps to reduce the *"needless or inadvertent release of persistent pesticides into the environment."* He said, *"Although the Department of Agriculture programs are carefully planned and carried out to insure maximum safety to man, animals, and our natural resources, the present concern over protection of our environment from contamination warrants a further review of our control operations."*

In another action, *Home Garden*, a national horticultural magazine for consumers, has announced a ban effective with the September issue on advertisements for products containing DDT and certain other chlorinated hydrocarbon insecticides which could have harmful effects on man and his environment.

DDT Goes to Trial in Madison

Charles F. Wurster

The hearings in Madison, Wisconsin, on the question of the persisting residues of DDT becoming one of the world's most serious pollution problems, involved 27 days of testimony from 32 witnesses, filled nearly 3000 pages of transcript and included 208 exhibits. The hearings adjourned 21 May 1969 and a decision on the use of the pesticide in Wisconsin is pending. (BioScience 19, no. 9, p. 809-813)

Scientists Focus on an Environmental Problem

An increasing groundswell of involvement by scientists in issues of social relevance has become evident recently, especially where matters of environmental quality are concerned. The feeling seems to be—act now, for tomorrow may be too late. The DDT problem has become a focus of this increased activity, and seldom has the confrontation been more intense and involved more scientists than at the DDT hearings before the Department of Natural Resources in Madison, Wisconsin.

It is not accidental that DDT is the focus or spearhead of scientists' concern. Mounting evidence now indicates that the persisting residues of DDT have become one of the world's most serious pollution problems, affecting many species of nontarget organisms on a global basis. Furthermore, the scientific literature on DDT is now voluminous, and it is probable that no man-made environmental contamination problem can be quite so thoroughly documented.

The author is an assistant professor of biological sciences at the State University of New York at Stony Brook, and Chairman of the Scientists Advisory Committee of the Environmental Defense Fund, Inc. His research has involved studies on the effects of DDT on nontarget organisms.
The section of this article on the history of the Environmental Defense Fund has been omitted because of space limitations. This section will be included in a similar article by Dr. Wurster which will appear in the September issue of *Audubon Magazine*, and is described by Carter (1967, 1969).

DDT is well known from its glamorous past, having undoubtedly saved millions of lives during and shortly after World War II when alternative insecticides were unavailable. The chemical is staunchly defended by some segments of the agricultural community and the chemical industry, and by a dwindling number of scientists. They insist that there is a continuing need for DDT in agriculture and public health and express concern that if DDT is banned, the use of several other chlorinated hydrocarbons with similar properties will also be curtailed.

Moreover, a battle over control is one of the central issues. *Who* is going to control pesticides? Historically, complete control has been in the hands of agriculture. Now that certain pesticides are affecting nonagricultural interests and values, a growing chorus of other voices, including the environmental science community, demand to be heard. If the environmentalists win on DDT, they will achieve, and probably retain in other environmental issues, a level of authority they have never had before. In a sense, then, much more is at stake than DDT.

The Judicial Process

Although formally the Madison hearings must be classified as administrative, conduct of the proceedings was quasi-judicial. Filling a role equivalent to that of judge was the Chief Hearing Examiner, Maurice Van Susteren, a man of considerable capabilities in both law and science. The rules of evidence—relevance, materiality, and competence of testimony—were rigorously observed, and the eventual decision, whichever way it goes, is appealable to higher courts. Testimony could be presented without time limitation, all of it subject to cross-examination.

Cross-examination played a vital role in the Madison story. Conducted by a skillful attorney who understands science in addition to trial law, cross-examination can separate fact from fiction by "sweating the truth out of a witness." Without well-directed, expert cross-examination, all kinds of testimony may sound about equally valid; this is often true when legislative committees lack the scientific expertise to ask the relevant questions or to judge the competence of a witness. Their reports and resultant legislation may therefore fail to grasp the problem and be ambivalent in response. Cross-examination is the acid test of relevance and competence.

Victor J. Yannacone, Jr., the Environmental Defense Fund's (EDF) trial attorney, has an impressive grasp of scientific material, especially the environmental sciences. His cross-examination is usually aggressive and may be devastating where a witness takes a position that is scientifically weak. Cross-examination can be a very rough business. It gives prospective witnesses an incentive to be well prepared and confident of the validity of their testimony—or not testify at all. The specter of cross-examination thereby "selects" witnesses in advance, tending to separate fact from fiction.

EDF had available far more scientists without fee than it could present as witnesses. The Task Force for DDT of the National Agricultural Chemicals Association (NACA), on the other hand, apparently had difficulty finding sufficient scientists to uphold its position. It seems curious that many of those who continue to promote the use of DDT, claiming that it is *not* a serious environmental hazard, did not appear at the well-publicized Madison hearings. One would think they would have welcomed the opportunity to test the validity of their opinions in a forum where each scientist was afforded ample time to present scientific evidence for his position.

The scene of action over DDT among the Citizens Natural Resources Association (CNRA) and EDF people in Madison was by no means restricted to the hearing room. To the contrary, a vigorous, occasionally heated scientific dialogue usually involving a considerable number of scientists and several attorneys occupied all mealtimes and many evenings. The value of these sessions cannot be overstated. Scientific positions were tested and eliminated if weak; stronger positions were discovered, tested again, and sometimes demolished, then reconstructed. Only after the stimulation and probing of such an interdisciplinary group did scientists take the stand. They sometimes remarked later that this intensive dialogue was more rigorous than the cross-examination of the opposing attorneys.

The "Environmental Scientist"

The Madison case was remarkable for the great diversity of disciplines that played a role in its organization and presentation. Not only were most of the biological and medical sciences involved, but the physical sciences including physics, chemistry, mathematics, meteorology, and even engineering played vital roles as well. The EDF approach was strongly problem-oriented, thereby requiring interdisciplinary co-operation, rather than being restricted to an orientation by discipline. The testimony by Orie L. Loucks, plant ecologist from the University of Wisconsin (UW) and the final witness of the hearing, was preceded by a week of intensive work involving mathematicians, various ecologists, systems analysts, mechanical and control engineers, and the searching probes of Yannacone. The result was a summary of the available data on DDT transport, uptake, and metabolism in the form of a complex, modern systems analysis that left the Task Force for DDT incapable of critical cross-examination.

A basic element of the EDF position has been to emphasize the complexity of the environmental sciences and the need for interdisciplinary cooperation, then to demonstrate through cross-examination that the DDT proponents are narrow specialists out of touch with the rest of the scientific community. The "environmental scientist" capable of evaluating the *total environment*, rather than its separate components, was repeatedly defined as a scientist with his own specialties who is in constant communication with other scientists of different disciplines.

Testimony Begins

The hearing was opened in the State Capitol Building on 2 December 1968, with Senator Gaylord Nelson as the first witness for the petitioners. Louis A. McLean, the attorney originally representing the Task Force for DDT of the NACA who was later replaced by Willard D. Stafford, was then called as an adverse witness by Yannacone for questioning about his article in *BioScience* (September, 1967, p. 613) in which he gave his views on "anti-pesticide people" and their preoccupation with "sexual potency" and various forms of "quackery." Another adverse witness called early was Ellsworth H. Fisher, entomologist from UW and an outspoken DDT advocate. The intent was to put into the record a rough outline of the opposition's case to come; "you can't shoot at empty air," says Yannacone.

The initial EDF approach was to describe the "Wisconsin Regional Ecosystem" with emphasis on its interconnections and interrelationships, prior to presenting any testimony about DDT and its role when injected into that system. Hugh H. Iltis, UW botanist, described these plant-animal-environment relationships in detail. Loucks followed with testimony about air currents, weather fronts, long distance pollen dispersal, and nutrient cycling, again emphasizing how events in one region could have distant effects. Iltis and Loucks both showed the inseparability and interrelationships between agricultural regions and the overall ecosystem of which they are but a part. They thus set the stage for the final systems analysis summation on the last day of the hearings 6 months later.

With these ecological relationships established, I took the stand to describe the physical, chemical, and biological properties of DDT (Wurster, 1969), testifying that DDT combines in a single molecule the properties of broad biological activity, chemical stability, mobility, and solubility characteristics that cause it to be accumulated by living organisms, thus presenting an environmental situation that is almost unique among major pollutants. DDT not only enters food chains from the inorganic environment, but it is increasingly concentrated toward the top of food chains, thereby posing a particular threat to carnivores (Woodwell et al., 1967).

My direct testimony took about one hour, but was followed by nearly 3 full days of wide-ranging cross-examination by McLean. Many aspects of the pesticide controversy were probed, as well as my background and that of EDF. All of the expected arguments that were used 10-20 years ago against scientists who first reported environmental degradation by DDT were raised, but it was surprising to hear almost none that were new. EDF soon learned that the Task Force was not fully prepared with regard to the current scientific literature on DDT.

Early in the trial, there were repeated attempts to equate DDT with pesticides in general to create the impression that EDF was an "anti-pesticide" organization. References to "pesticides" were invariably met by objections from Yannacone. These were usually sustained by Van Susteren, who reminded McLean to confine his questioning to "DDT," not "pesticides."

In its presentation, EDF emphasized those effects of DDT that affect a whole species and are world-wide in magnitude, rather than the more local, though more spectacular, fish or bird kills. The use of DDT for attempted control of Dutch elm disease, however, has long been of special concern in the Middle West, and the dramatic bird mortality caused by this usage was described by George J. Wallace (1959), zoologist from Michigan State University, and Joseph J. Hickey, wildlife ecologist from UW, both among the early scientists reporting such damage. William Gusey of Shell Chemical Company and a DDT Task Force witness said that winter spraying of elms, before birds are in the area, eliminates robin mortality, but Wallace contradicted this by telling how robins continued to die for years after the last DDT spraying because they ate contaminated earthworms.

Wallace described the tremors he observed in birds dying of DDT poisoning. He was followed on the stand by Alan Steinbach, neurophysiologist from Albert Einstein College of Medicine, who gave a comprehensive description of the mechanisms of nerve transmission and their disturbance by DDT. He discussed the Hodgkin-Huxley equation and the effects of various toxins on transmission of the nerve impulse, indicating that the effects of DDT are irreversible as compared with these other toxins. "The known mechanism of action of DDT on nerves . . . can account for . . . the observations outlined by Dr. Wallace, both today and in his earlier papers," Steinbach concluded.

During the early weeks of the hearing there was objection by McLean to the term "biocide" when used by several scientists in reference to DDT. Included among the pro-DDT witnesses, however, was Taft A. Pierce of the Orkin Exterminating Company, who testified that DDT is especially useful for killing mice and bats. Pierce said that without DDT, he would have to use "poison" more frequently to control mice.

The almost ubiquitous distribution of DDT residues in the world environment was the subject of testimony by many scientists. Robert W. Risebrough, molecular biologist from the Institute of Marine Resources, University of California at Berkeley, called DDT and its metabolites "the most abundant synthetic pollutant(s) in the global ecosystem" (Risebrough et al., 1968b). He described their dispersal by currents of air and water, and his discovery of these materials in the air over Barbados (Risebrough et al., 1968a). His analyses showed oceanic fish and birds from various parts of the Pacific to be contaminated, often to "very high levels," and Hickey gave analytical data on residues of DDE (a DDT metabolite) in the Lake Michigan ecosystem showing exceptional contamination of many organisms (Hickey, et al., 1966). "Lake Michigan, in spite of its size, "Hickey said, "is one of the most polluted (with DDE) lakes in the world."

DDT, Thin Eggshells, and Reduced Reproduction

Enzyme induction by chlorinated hydrocarbons, including DDT, was for several years primarily of interest to biochemists, but recently environmental scientists have discovered that the phenomenon may be of enormous ecological significance. Richard M. Welch, biochemical pharmacologist from Burroughs Wellcome and Company, told how DDT induces the synthesis of hepatic microsomal enzymes in a variety of test animals (Conney, 1967). These liver enzymes have a broad spectrum of activity that includes the ability to hydroxylate the steroid sex hormones testosterone, progesterone, and estrogen when DDT is administered to test animals at concentrations described by other scientists as common in the environment. Welch told how, in the course of these experiments, he and his colleagues had accidentally discovered that DDT can itself also function as an active estrogen (Welch et. al., 1969), thus presenting yet another mechanism of action for the material.

It took the testimony of several scientists to clarify the environmental importance of enzyme induction. Earlier, Risebrough and I had pointed out that estrogens affect calcium metabolism and eggshell formation in birds, and that elevated estrogen metabolism caused by DDT-induced enzymes could depress estrogen levels and result in birds laying eggs with thin shells (Peakall, 1967; Wurster, 1968). Hickey (1969) then described the great population collapses among several species of carnivorous birds, particularly the peregrine falcon, on two continents during the past 15 years.

Characteristic of these population declines, Hickey related, was reproductive failure associated with various symptoms, especially egg breakage, suggesting disturbed calcium metabolism in these birds. The mystery has only been solved since 1967 through the close collaboration and discoveries of a diverse team of environmental scientists. Hickey told how Ratcliff (1967), he, and Anderson (1968) found that the eggshell thickness for these species, as measured in museum collections, had been stable for half a century, then underwent a sudden reduction simultaneously in Europe and North America during the late 1940's—shortly after the introduction of DDT into the world environment. He also pointed out that there was a statistically significant inverse correlation between concentrations of DDE in herring gull eggs and the thickness of their eggshells.

If this remarkable pyramid of circumstantial evidence involves a causal relationship between DDT, thin eggshells, and diminished reproductive success, then controlled feeding experiments with DDT or its metabolites should produce these effects. Exactly such experiments were described by Lucille F. Stickel, Patuxent Wildlife Research Center, Bureau of Sport Fisheries and Wildlife, U.S. Department of Interior, testifying only a few days after their completion. Groups of mallards were fed three parts per million (ppm) of DDE. Eggs from ducks receiving DDE, when compared with controls, had shells that were 13.5% thinner, were cracked or broken six times as often, and produced less than half as many healthy ducklings. DDT gave results that were comparable to those of DDE, but DDD did not affect shell thickness at this dosage (Heath et al., 1969).

In another experiment, kestrels, close relatives of the peregrine falcon, were fed 2 ppm of DDT plus 1/3 ppm of dieldrin. Dosed hawks laid eggs with shells that were 15% thinner than the controls, and their reproductive success was impaired (Porter and Wiemeyer, 1969). Although probably the shortest, Stickel's testimony was among the most important of the hearing because it established that thin eggshells and reduced reproduction among carnivorous birds are directly caused by residues of DDT at current environmental levels.

The Task Force for DDT made but one attempt to rebut the avian reproduction evidence. Frank L. Cherms, Jr., UW Department of Poultry Science, told of his experiments with Japanese quail, where 200 ppm of DDT in the diet had caused no effect on eggshell thickness or reproductive rate. He also described a variety of factors that can affect shell thickness, including heredity, nutrition, diseases, temperature, and humidity; "if you frighten them," he added, "you can scare the shell out of them." Under cross-examination, however, Cherms agreed that he was an expert on poultry but was not competent to discuss wild birds, their eggshell formation, or reproduction. Earlier, Stickel had pointed out that this quail lay eggs continuously throughout the year and differs from wild carnivorous birds that lay only in the spring. Robert L. Rudd of the Department of Zoology, University of California at Davis, also indicated that graminivorous pheasants could not be compared physiologically with carnivores at the top of a food chain.

Evidence that DDT reduces reproduction in fish, presented by Kenneth J. Macek of the Fish-Pesticide Research Laboratory (Columbia, Mo.), U. S. Department of Interior, was not contested by the DDT Task Force. Macek (1968a) described experiments in which 1 mg/kg/week of DDT in the diet of brook trout did not kill any adult fish, but caused increased mortality of fry from the residues stored in the egg yolk. He said that DDT residues accumulated in his experimental fish were comparable to those in fish from several freshwater lakes, including the Coho salmon from Lake Michigan, where abnormal fry mortality has been occurring. Risebrough gave similar analyses for fish from the Pacific Ocean, suggesting that important marine fisheries are threatened. In another experiment, Macek (1968b) found susceptibility to stress, as measured by associated mortality, increased from 1.2% in control fish to 88% in those fed diets containing 2 mg/kg/week of DDT.

DDT and Human Health

The original petition for the Wisconsin hearings made no mention of a relationship between DDT and human health, but the subject inevitably became the center of much testimony. Wayland J. Hayes, Jr., toxicologist from Vanderbilt University, said that DDT was "absolutely safe" for the human population at current exposure, based on studies of men exposed to much higher levels of DDT who showed "no clinical symptoms" (Hayes et al., 1956; Laws et al., 1967). Hayes' conclusion, however, was founded on tests for gross nervous system disorders and the assumption that DDT is a nerve poison, while other mechanisms, such as the enzyme effects described by several witnesses, were not investigated. He admitted that many biochemical and other sophisticated tests had not been made on these subjects, and that women, children, and infants had not been studied, yet he remained firm on his "absolutely safe" opinion.

The testimony of several scientists did not agree with Hayes'. Theodore L. Goodfriend of the UW School of Medicine said that "one cannot conclude that DDT is absolutely safe for human use." He described a variety of possible hormonal effects that have not been investigated and suggested they should be, since DDT is known to interfere with endocrine systems. Earlier, Welch had indicated that DDT at concentrations currently found in human fat was associated with elevated levels of steroid hormone hydroxylases in rats, and "if one can extrapolate data from animals to man, then one would say that a change in these liver enzymes probably does occur in man." He said that a controlled experiment had not been done in man, but an uncontrolled experiment "should not be done on the worldwide population." Hayes admitted he had done no such liver tests.

On 27 March 1969, Sweden announced a moratorium on several chlorinated hydrocarbon insecticides, including DDT, that was reached following the extensive literature research of a team of scientists—the Working Group on Environmental Toxicology, Ecological Research Committee of the Swedish Natural Science Research Council. Chairman of the scientists' team was Göran Löfroth of the Royal University of Stockholm, a surprise witness at the hearings appearing by invitation of the Public Intervenor and Assistant General of Wisconsin, Robert B. McConnell.

Löfroth discussed the ubiquitous world-wide distribution of DDT residues in human tissues and mentioned especially its presence in human milk (Curley and Kimbrough, 1969). He indicated that women excrete a higher proportion of ingested DDT into their milk than do cows, and that nursing infants receive about twice the maximum daily intake of DDT-compounds recommended by the World Health Organization.

Löfroth agreed with other scientists that the safety of DDT had not been demonstrated and that its use in the environment should therefore be discontinued. He also quoted publications suggesting that DDT interferes with normal body biochemistry, that it causes tumors in mice (Kemeny and Tarjan, 1966), and that there is a correlation between higher than average residues of DDT and the frequency of human deaths from various disorders including liver cancer (Deichmann and Radomski, 1968).

Possible PCB Interference

As an important part of its defense, the Task Force for DDT tried to demonstrate that analyses by environmental scientists were in error because there was interference from polychlorinated biphenyls (PCB's), industrial compounds that are also widespread in the environment. In an apparent attempt to cast doubt on the analyses of several scientists who had testified on behalf of EDF, the DDT Task Force presented Francis B. Coon, Head of the Chemistry Department of the Wisconsin Alumni Research Foundation (WARF) where many of these analyses had been performed, who testified that there was interference with DDT and DDD, and that he could not be certain there was no interference with DDE.

Since Hickey had presented only DDE analyses in his testimony because of the interference problem at WARF with DDT and DDD, the status of DDE became the object of much testimony. During a highly technical, day-long cross-examination by Yannacone, Coon changed several positions he had taken on direct examination, finally admitting that there was no significant PCB interference with the DDE analyses.

Further testimony on analytical techniques and PCB's was presented by Paul E. Porter of the Shell Development Company, who said that DDT, DDD, and DDE can be distinguished in the presence of PCB's using gas chromatography. Porter also expanded on earlier testimony about physical and chemical properties of DDT, its degradation, and its transport mechanisms. EDF considered Porter's testimony to be competent, accurate, and in no conflict with the petitioner's position; therefore he was not cross-examined.

Stafford also called Risebrough as an adverse witness for additional questioning about PCB's and their role in the thin eggshell phenomenon (Risebrough et al., 1968b), even though he had been cross-examined for 3 days by McLean in December. Risebrough, who attended almost all sessions of the hearing, told how DDE, and not PCB's, is largely responsible for the widespread occurrence of thin eggshells among carnivorous birds. He also used the opportunity to present new evidence on the almost complete reproductive failure early in 1969 of the brown pelicans in California, whose eggs collapsed when the birds tried to incubate them.

Appearing on behalf of the U. S. Department of Agriculture (USDA) was Harry W. Hays, Director of the Pesticides Regulation Division. Hays described USDA's procedures for the registration of a pesticide, indicating that his division within USDA has complete responsibility for registration, and that other federal agencies, including the Departments of Health, Education, and Welfare, and Interior, have only an advisory capacity without authority. He said that the chemical company applying for the registration supplies the data to USDA, but under cross-examination Hays revealed that USDA makes no independent check of these data, other than for internal consistency.

Yannacone led Hays through a complicated tangle of bureaucratic processes and responsibilities within USDA. Hays said that during the past few years there have been new registrations for DDT, but that new data on sublethal effects on animals or mobility of DDT had not been required of the applicants. Yannacone repeatedly asked about procedures for cancelling the registration of DDT, and was told that only the Secretary of Agriculture could initiate cancellation proceedings.

Another DDT proponent was Samuel Rotrosen, president of Montrose Chemical Corporation, the largest DDT maker in the United States. Rotrosen said that the Montrose plant in Los Angeles makes about half of the U. S. total, and that the other chemical companies manufacturing DDT are Diamond-Shamrock, Olin-Mathiessen, Allied, and Lebanon. Of 114 million lb. made last year, about two-thirds was exported, and Rotrosen described its various domestic and foreign uses. In carload lots of 60,000 lb., DDT costs 17¢ per lb., and total production last year was valued at about $20 million.

Integrated Control of Insect Pests

Testifying on behalf of the Task Force for DDT were two entomologists, R. T. Chapman of the UW Department of Entomology and Bailey B. Pepper of the Department of Entomology of Rutgers University. Chapman said that DDT is still needed to prevent insect damage to certain crops, especially cabbage and carrots in Wisconsin. Pepper also mentioned several agricultural needs for DDT, as well as the threat of mosquito-borne encephalitis. Under cross-examination by Yannacone, Pepper agreed that malathion is an alternative for mosquito control and DDT is not recommended on salt marshes.

In its presentation, EDF emphasized not only the deleterious effects of DDT, but alternative insect control techniques as well. Three top scientists in the sophisticated field of integrated control, the blending of biological and chemical insect control techniques in an integrated system, testified at length on the effectiveness and desirability of this approach to agricultural pest problems.

Robert van den Bosch, entomologist in the Division of Biological Control, University of California at Berkeley, described ways in which an agro-ecosystem can be manipulated "to manage pest populations so that they do not cause economic loss" (Smith and van den Bosch, 1967). He discussed the role of entomophagous insects (insects that are parasites and predators of other insects) in controlling potential pest species and pointed out that DDT often causes eruptions of the pest populations by elimination of these entomophagous insects.

Van den Bosch told how for many years he had recommended DDT for some purposes, but that more recent knowledge of its enormous ecological impact had caused him to discontinue these recommendations. He described DDT as an "ecologically crude material . . . developed by chemists and toxicologists . . . with no ecological thought whatsoever," and "exploited largely by people who were thinking in terms of the economics . . . . Most entomologists eagerly seized" the material "totally ignorant of the . . . ecological implications" of its use; of those implications, he said, "I'm scared."

Testifying on 21 May, Paul DeBach of the Department of Biological Control, University of California at Riverside, and Donald A. Chant, Chairman of the Zoology Department, University of Toronto, reiterated van den Bosch's position. DeBach (1964) described DDT as a highly disruptive material in an agro-ecosystem and told how it causes outbreaks of mites and scale insects by killing their natural enemies. Chant (1966) discussed the concept of economic threshold, pointing out that insecticides are often used when the pest population is below a level of economic damage, or even totally absent from the area. "DDT," he said, "has no place in integrated control."

DDT in the Ecosystem

Summation testimony was given both by Loucks and Rudd. Following his systems analysis presentation, Loucks concluded that DDT concentrations at higher trophic levels in the Wisconsin ecosystem can be expected to increase, with further decreases in numbers of important predator species leading to instability and degradation of the ecosystem.

Rudd (1964) summed up the problem by saying that pest control operations too often have been restricted to a consideration only of a particular pest and a particular crop. "The pest control operator, once the applications have been made, pretty well forgets the problem . . . . The ecologist, on the other hand, is concerned with entire . . . ecosystems. He has no particular restrictions." Both Rudd and Loucks said that DDT fits the definition of a pollutant in Wisconsin law since it is "deleterious to fish, bird, animal or plant life."

The Madison DDT hearings involved 27 days of testimony from 32 witnesses, filled nearly 3000 pages of transcript, included 208 exhibits, and adjourned on 21 May 1969, nearly 6 months after they had begun. The decision is pending.

References

Carter, L. J. 1967. Environmental pollution: Scientists go to court. *Science,* **158:** 1552-1556.

──────. 1969. DDT: The critics attempt to ban its use in Wisconsin. *Science,* **163:** 548-551.

Chant, D. A. 1966. Integrated control systems. *Scientific Aspects of Pest Control,* Nat. Acad. Sci.—Nat. Res. Council, Publ. 1402, p. 193-218.

Conney, A. H. 1967. Pharmacological implications of microsomal enzyme induction. *Pharmacol. Rev.,* **19:** 317-366.

Curley, A., and R. Kimbrough. 1969. Chlorinated hydrocarbon insecticides in plasma and milk of pregnant and lactating women. *Arch. Environ. Health,* **18:** 156-164.

DeBach, Paul. 1964. *Biological Control of Insect Pests and Weeds.* Reinhold Publishing Corp., New York, 844 p.

Deichmann, W. B., and J. L. Radomski. 1968. Retention of pesticides in human adipose tissue—preliminary report, *Ind. Med. Surg.,* **37:** 218-219.

Hayes, W. J., Jr., W. F. Durham, and C. Cueto, Jr. 1956. The effect of known repeated oral doses of chlorophenothane (DDT) in man. *J. Amer. Med. Assoc.,* **162:** 890-897.

Heath, R. G., J. W. Spann, and J. F. Kreitzer. 1969. Marked DDE impairment of mallard reproduction in controlled studies. *Nature* (in press).

Hickey, J. J. (ed.) 1969. *Peregrine Falcon Populations: Their Biology and Decline.* The University of Wisconsin Press, Madison, 596 p.

Hickey, J. J., and D. W. Anderson. 1968. Chlorinated hydrocarbons and egg-shell changes in raptorial and fish-eating birds. *Science,* **162:** 271-273.

Hickey, J. J., J. A. Keith, and F. B. Coon. 1966. An exploration of pesticides in a Lake Michigan ecosystem. *J. Appl. Ecol.,* **3** (Suppl.): 141-154.

Kemeny, T., and R. Tarian. 1966. Investigations on the effects of chronically administered small amounts of DDT in mice. *Experientia,* **22:** 748-749.

Laws, E. R., Jr., A. Curley, and F. J. Biros. 1967. Men with intensive occupational exposure to DDT. *Arch. Environ. Health,* **15:** 766-775.

Macek, K. J. 1968a. Reproduction in brook trout *(Salvelinus fontinalis)* fed sublethal concentrations of DDT. *J. Fish. Res. Board Can.,* **25:** 1787-1796.

──────. 1968b. Growth and resistance to stress in brook trout fed sublethal levels of DDT. *J. Fish. Res. Board Can.,* **25:** 2443-2451.

Peakall, D. B. 1967. Pesticide-induced enzyme breakdown of steroids in birds. *Nature,* **216:** 505-506.

Porter, R. D., and S. N. Wiemeyer. 1969. Dieldrin and DDT: Effects on sparrow hawk eggshells and reproduction. *Science,* **165:** 199-200.

Ratcliffe, D. A. 1967. Decrease in eggshell weight in certain birds of prey. *Nature,* **215:** 208-210.

Risebrough, R. W. et al. 1968a. Pesticides: Transatlantic movements in the Northeast trades. *Science,* **159:** 1233-1236.

──────. 1968b. Polychlorinated biphenyls in the global ecosystem. *Nature,* **220:** 1098-1102.

Rudd, R. L. 1964. *Pesticides and the Living Landscape.* University of Wisconsin Press, Madison, 320 p.

Smith, R. F., and R. van den Bosch. 1967. Integrated control. *Pest Control* (Chap. 9), Academic Press, Inc., New York, p. 295-340.

Wallace, G. J. 1959. Insecticides and birds. *Audubon Mag.,* **61:** 10-12, 35.

Welch, R. M., W. Levin, and A. H. Conney. 1969. Estrogenic action of DDT and its analogs. *Toxicol. Appl. Pharmacol.,* **14:** 358-367.

Woodwell, G. M., C. F. Wurster, and P. A. Isaacson. 1967. DDT residues in an East Coast estuary: A case of biological concentration of a persistent insecticide. *Science,* **156:** 821-824.

Wurster, C. F. 4-6 June 1968. Chlorinated hydrocarbon insecticides and avian reproduction: How are they related? First Rochester Conference on Toxicity, University of Rochester.

──────. 1969. Chlorinated hydrocarbon insecticides and the world ecosystem. *Biol. Conserv.,* **1:** 123-129.

Radioactivity and Fallout: The Model Pollution

George M. Woodwell

This article attempts to summarize the broader scientific aspects of the problems associated with release of certain types of toxic wastes into the environment. The author examines the radiation-pollution problem and shows how it can be used to provide a solution for certain other analogous pollution problems, especially those of pesticides. (BioScience 19, no. 10, p. 884-887)

Ecologists are so thoroughly accustomed to playing Cassandra, predicting doom, that we hardly give credit for even major postponements of the day of doom. Doom has so many components these days that it is difficult to sort out which one is more important at any moment: the crisis of the dollar, the war, the cities, the pollution, the population, or the fact that the students are going to pot. What is encouraging and central to a symposium entitled "Challenge for Survival" is the fact that there is now a broad and growing consensus that accepts the simple truth that the size of the human population is a key and that the ultimate amelioration of these crises depends on establishing some sort of equilibrium between population and resources. Establishing any such equilibrium is difficult, almost hopelessly so, not only because of the difficulties of controlling the numbers of people but also because a burgeoning, and in some ways, malignant technology both increases certain types of resources and simultaneously destroys other essential ones. Thus an expanding technology offers ever cheaper power and transportation but also threatens to degrade the environment in diverse ways, one of the most important being with wastes that are biologically active.

The author is a member of the Biology Department, Brookhaven National Laboratory, Upton, New York 11973.
Research carried out at Brookhaven National Laboratory under the auspices of the U.S. Atomic Energy Commission.
Paper presented before the New York Botanical Garden-Rockefeller University Symposium "Challenge for Survival," held at the Rockefeller University, New York City, 25 April 1968.

My objective is to summarize the broader scientific aspects of the problems associated with release of certain types of toxic wastes into the environment. Atomic energy provides a model, unfinished, rough-hewn in many of its parts, polished in others, but overall a brilliant example of the marshalling of scientific and political talent on an international scale to mitigate a world-wide pollution problem. That problem has not been solved, but there is hope that it can be and there are many lessons from it. I propose to examine the radiation-pollution problem and to show how it can be used to contribute to a solution of certain other analogous pollution problems, especially those we have now with pesticides, which appear to be the world's most dangerous pollutants.

First, the attitudes that allowed worldwide contamination of the earth with radioactivity and which now allow other even more serious pollutions are important and still dominant, although weakening. The most important assumption that led to the problems with radioactivity is the assumption of dilution. Toxic materials released into the environment are widely assumed to be diluted to innocuousness. If there are local effects from the toxicity, they are transient; an abundant and vigorous nature repairing any damage within at most a few years—or the effects are accepted as small cost for technological progress. The assumption is practical and as long as the environment is very large in comparison with the quantity of the release, "dilution" appears to occur. At least the material "disappears."

So thoroughly ingrained is this philosophy that its corollary, the right to pollute, has become a second major philosophical and, for somewhat different reasons, legal assumption: we tend to require detailed scientific proof of direct, personal damage to man as a prerequisite for even considering restriction of any right to pollute.

On the basis of these two assumptions—dilution and the assumption of the right to pollute until proof of damage—we, man all over the world, are now well embarked on a program of releasing unmeasured quantities of different kinds of biologically active substances into the general environment. It is one of the spectacular contradictions of our time that in the age of science we should be entering blindly on a thousand, unplanned, uncontrolled, unmonitored, unguided, largely unrestrained, and totally unscientific experiments with the whole world as the subject and survival at hazard.

The assumption of dilution, so easy to make, so cheap, so comforting, so much a part of human nature, is a trap. Biologically active materials released into the biosphere travel in patterns that are surprisingly well known. A major contribution of atomic energy has been definition of these patterns, using as tracers the radioactivity in fallout from bomb tests. Intensive studies began in earnest only after the series of bomb tests in the Pacific in the spring of 1954 known as

"Castle." This was the series, some of you will recall, that included the test that dropped fallout on Rongelap Atoll, exposing its inhabitants, about 65, to an estimated 175 r (500 r is widely accepted as the mean lethal exposure for man). A Japanese fishing vessel, the *Lucky Dragon*, and its crew were also caught in the fallout; and for several months tuna caught in the Pacific and landed in Japan were sufficiently radioactive that authorities would not allow them to be sold. In the eyes of the world this series of misfortunes was "proof of damage" and frightening proof of a new-found capacity to degrade the environment in places far removed from those affected directly by the blast and the ionizing radiation accompanying the bomb. Some of the radioactivity from these tests is, of course, still circulating through the biosphere. It was this series of frightening events, publicized widely at the time and pursued doggedly over the years by many knowledgeable and deeply concerned scientists, that triggered sufficient public interest in the problems of world-wide pollution with radioactivity to mount a really significant international research program tracing these patterns (UNSCEAR, 1958, 1962, 1964) and that ultimately resulted in the treaty among the more civilized nations banning nuclear tests in the atmosphere.

From the research on man-made radioactivity we have learned these things, all relevant and even basic to the greater problems of pollution that we are just now beginning to recognize as a major "challenge to survival":

First, particulate matter, introduced into the lower atmosphere, enters air currents that move around the world in periods of 15 to 25 days in the middle latitudes, sometimes less.

Second, the half-time of residence (time for one-half of the material to be removed) of particulate matter carried in these currents ranges between a few days and a month, the same general range as the time to travel around the world, Thus, it is no surprise these days to measure the radioactive cloud produced by tests in central Asia on several successive trips around the world.

Third, such material tends to be removed from air and deposited on the ground by precipitation, more in the early precipitation in any storm than later.

Fourth, the patterns apply to any particulate matter entering the air currents of the troposphere. Some fraction of the pollen, for instance, that is released by plants close to the ground surface enters these patterns and is transported in air and deposited in precipitation (Gatz and Dingle, 1965).

Hardly less important, we know that certain radionuclides are accumulated from the environment into the tissues of plants and animals, where they may be concentrated in high degree, concentration factors of a hundred to a thousand-fold and higher being common. But the patterns of movement or radionuclides through ecological systems are not capable of simple generalization: each radionuclide travels its own peculiar path. Thus ^{137}Cs behaves in ways that are similar to K and tends to be accumulated in muscle with greatest accumulations in older organisms and in carnivores. The accumulation of ^{137}Cs in Eskimos through the lichen-caribou food chain is well known. However, ^{90}Sr is accumulated in bones and herbivores get most of it; ^{137}I, in thyroids; ^{55}Fe, in blood and elsewhere. Discovery of these patterns has required intensive research on each substance, both to trace its movement through the patterns of air and water circulation and to discover its pathways through living systems (Åberg and Hungate, 1967; Polikarpov, 1966).

But this is only part of the story. It documents the fact that dilution into an infinite environment is not a safe assumption, but what of the effects? Again the story is not a simple one, each substance presenting its own peculiar set of hazards. With ionizing radiation the problem seems reasonably straightforward although it has never seemed very straightforward to those charged with developing standards of safety. There is general agreement that the principal hazard is a direct hazard to man through damage to the genetic material, "mutations," which are, for the practical purposes of this discussion, all deleterious. Man is most vulnerable because radiation causes an increase in the frequency of deleterious mutations, in the jargon, an increase in the "genetic load," by adding to the numbers of genetically determined unfortunates. If man is protected from this hazard, levels of man-made radiation in nature will almost certainly be so low as to have no significant effects on other organisms, because in these species, unlike man, genetic unfortunates are removed by selection. Increasing mutation rates will probably not increase rates of evolution as so often assumed. Thus the hazard of ionizing radiation, released from whatever source, is first a direct hazard to man; this is not so for many other toxic substances as we shall see in a moment.

But again, there is no simple answer to the question of how much radiation is "safe." It would be very convenient if there were some threshold below which ionizing radiation has no effect. A considerable weight of evidence suggests that there is no threshold for production of mutations; even very low exposures increase mutation frequencies slightly. On the other hand, there is clear benefit from radiation exposures in medicine, and it is not possible to enjoy the benefits of nuclear power without some small increase in radiation exposure for workers in the plant. We must arrive at a compromise in exposing people directly, limiting direct exposures around nuclear plants systematically and vigorously to those well below levels that increase genetic hazards appreciably. But the problems raised by the cycling of radioactive wastes are different. It is more difficult to anticipate exposure, more difficult to make even a reasonable guess as to how much of any nuclide will appear in human tissues. And, due to the movement of air and water around the world, the problem for long-lived materials suddenly becomes not merely a local one, confined to the Hudson River Estuary, the Columbia River, the Irish Sea, or the Bay of Biscay, all water bodies now receiving appreciable radioactive wastes, but a world-wide one. Each increment of waste enters a world-wide pool of that material. Once we recognize that many pollution problems are world problems, not local ones, then we can approach them systematically, estimating the totals that we are willing to have in the biosphere at one time. Radioactive materials decay at some constant rate, often defined as "half-life," the time for half the activity to be lost. Thus there is a constant rate of removal of any radioactive substance from the environment due simply to physical decay. If we know that a quantity (A) of a substance is released into the environment regularly and that at the end of time (t) the fraction $\frac{A_t}{A_o} = R$ remains, then the amount (S) present after n units of time will be:

$$S = A\frac{(1-R^n)}{1-r} \quad (1)$$

R is related to half-life ($t_{\frac{1}{2}}$) by:

$$R = e^{-0.693\frac{t}{t_{1/2}}} \quad (2)$$

where t is the period for which the retention rate is derived and t and $t_{\frac{1}{2}}$ are in the same units. The equilibrium concentration will be given by assuming that $n \to \infty$, when

$$S = \frac{A}{1-R} \quad (3)$$

Thus the equilibrium concentration of something with a half-life in the range of 1 year will be about twice the annual release. If the half-life is 10 years, the equilibrium will be about 15 times the annual release. A material with a half-life of 30 years, such as ^{137}Cs, will achieve an equilibrium in the environment that is about 50 times the annual release. If the substance is sequestered in forms that make it unavailable for circulation through living systems, then the equilibrium calculated on the basis of physical half-life will simply state the maximum that could be circulating. The actual amount will depend on the efficiency of the mechanism that sequesters the radioactivity. For example, ^{137}Cs circulates freely in biological systems, but tends to be fixed in certain micaceous clays and effectively removed from further circulation. The rate of fixation in clays, however, is not easily estimated.

What is clear is that there is need to relate the input of toxic wastes to their world-wide equilibriums and these equilibriums to the changes such concentrations will induce in the biosphere. With ionizing radiation man is more vulnerable than his environment. The most serious radiation hazards will ultimately arise from gases such as tritium and ^{85}Kr which are long-lived, produced in appreciable quantities, and are difficult to contain. The problems may be aggravated as atomic energy becomes more widely used by industry because the interests of long-term safety and industrial profits are not always coincident. And there will be other proposals such as the one to build a new Isthmian canal with nuclear explosives; and there will be continuing military pressure to test bombs in the atmosphere again. There is much more evidence, however, to suggest that we know how to control these problems than there is that we recognize and can control other analogous ones.

There are probably large quantities of many different kinds of metabolites of civilization circulating in the biosphere right now, substances that are released in large quantities, are biologically active, have long half-lives, and therefore present problems that are analogous to those presented by ionizing radiation. The persistent pesticides, however, seem to be by far the greatest problem, but not, strangely enough, by poisoning man. With pesticides there is a hazard to man, but we have taken pains to insulate man's food chain by elaborate regulations. By and large these regulations are effective, although not always (*New York Times*, Sunday, 7 April 1968). What has happened is that we have at least temporarily protected man, but we have allowed virtually unlimited uses of long-lived pesticides in any place that will not contaminate human food produced in agriculture. The result has been the accumulation in the biosphere of concentrations of persistent pesticides, especially DDT, that are quite clearly degrading ecological systems all over the world. The extent of the changes are far from clear. What is clear is that certain carnivorous and scavenging birds the world over are suffering rapid reductions in reproductive capacity that seem clearly related to pesticide burdens. This applies even to oceanic birds such as the Bermuda petrel that do not ever come into contact with man or with sprayed areas (Wurster and Wingate, 1968). But there is ample reason to believe that many groups in addition to birds are affected, including oceanic fisheries and perhaps even phytoplankton, the basis of all oceanic food chains (Wurster, 1968). The trend, if allowed to continue, will follow the pattern set by eutrophication of Lake Erie and numerous other smaller lakes, now being followed rapidly by Lake Michigan. While we may be able to afford to lose lakes in this way, we cannot afford to lose the oceans.

It is hardly surprising that we have an acute problem with pesticides when one recognizes that probably very close to 100% of the world production of DDT is distributed in places where it can move freely through the various cycles of the biosphere. There is good evidence to support the assumption that the chemical degradation of pesticides follows the same pattern as radioactive decay: the amount decaying is proportional to the concentration (Hamaker, 1966). This means that we can use the same arithmetic to estimate the total quantity that will be circulating in the biosphere at equilibrium if we know the half-life and the rate of release.

The half-life of pesticides in nature is not easily estimated. Studies of agriculture soils indicate a half-time for disappearance of residues of DDT in the range of 2-4 years. But undoubtedly this reflects several types of losses including erosion, vaporization, co-distillation with water, and leaching into the water table as well as chemical degradation. When organic soils are present, residues tend to remain for many years (Woodwell and Martin, 1964; and others). An estimate of half-life of 10 years for DDT residues in biological systems seems minimum; their persistence may be considerably longer.

The annual production of pesticides in the world appears not to be tabulated. The U.S. Tariff Commission reports U.S. production of DDT. Between 1957 and 1967 production ranged between 99 and 179 million pounds, with the highest in 1963. In 1967 production was about 103 million pounds. The mean for the 11-year period was about 147 million pounds. If U.S. production is 75% of world production, then we might assume with some justification that the world equilibrium would be based on an annual release of 200 million lb. into the biosphere. The total amount of DDT residues circulating in the biosphere would then be about 3 billion lb., but this equilibrium would be approached only after 75 years. If we had used DDT at this high rate since 1946, we would now have about 1.5 billion lb. in the biosphere or about one-half the total we would have if the residues came to equilibrium under these conditions. Thus we can expect far greater changes in the world's biota than we have seen so far if we continue using these long-lived pesticides.

My colleagues will be quick to point out that DDT production in the United States has dropped in the past year, and there is some evidence of a downward trend in a period when total use of pesticides is increasing abruptly. With more than a billion pounds of DDT now cycling in the biosphere, and with abundant evidence of effects, there can hardly be a downward trend in use that is steep enough to avoid irreparable changes in the earth's biota, changes that can only be deleterious to our long-term interests in survival.

These observations simply point once more to the fact that the world is now small and we must tidy it up if we intend to continue using it for very long. The history of ionizing radiation is an example of a pollution whose hazards we have appraised and whose sources we have controlled. Pesticides are an example of the opposite: failure to appraise the hazards; and now so much political and industrial power supporting the status quo that we may not be able to control it before we have lost an important fraction of the earth's biota, driving the whole earth a significant step down the path that Lake Erie has followed.

To return to my original point, survival demands control of population, but it also demands some much more stringent and unpopular limitations on technology. It is true that there is some ultimate limit on the amount of power that can be produced by atomic energy without exposing man to unacceptable levels of radioactivity. It is also true that there is a limit on the amount of persistent pesticides we can tolerate circulating through the biosphere. Testing these limits as we are doing at the moment is a strange and dangerous game, suggesting that we have not yet learned that the *Challenge* is for *Survival*.

References

Åberg, B., and F. P. Hungate (eds.). *Radioecological Concentration Processes* (Proc. Intern. Symp., Stockholm, 25-29 April 1966), Pergamon Press, London, 1040 p.

Gatz, D. F., and A. N. Dingle. 1965. Air cleansing by convective storms. In: *Radioactive Fallout from Nuclear Weapons Tests*, A. W. Klement, Jr. (ed.), Div. Tech. Information, Oak Ridge, Tenn., p. 566-581.

Hamaker, J. W. 1966. Mathematical prediction of cumulative levels of pesticides in soil. In: *Organic Pesticides in the Environment*, Advances in Chemistry Series, No. 60, R. F. Gould (ed.), American Chemical Society, Washington, D.C., p. 122-145.

New York Times. Sunday, 7 April 1968. Montana Dairies Shut in Poisoning.

Polikarpov, G. C. 1966. *Radioecology of Aquatic Organisms* (translated from the Russian by Scripta Technica, Ltd.), English translation, V. Schultz and A. W. Klement, Jr. (eds.), Reinhold, New York, 314 p.

Report of the United Nations Scientific Committee on the Effects of Atomic Radiation. 1958. General Assembly, Official Records: 13th Session, Suppl. No. 17.

Report of the United Nations Scientific Committee on the Effects of Atomic Radiation. 1962. General Assembly, Official Records: 17th Session, Suppl. No. 16.

Report of the United Nations Scientific Committee on the Effects of Atomic Radiation. 1964. General Assembly, Official Records: 19th Session, Suppl. No. 14.

Woodwell, G. M. and F. T. Martin. 1964. Persistence of DDT in soils of heavily sprayed forest stands. *Science,* 145: 481-483.

Wurster, C. F. 1968. DDT reduces photosynthesis by marine phytoplankton. *Science,* 159: 1474-1475.

Wurster, C. F. and D. B. Wingate. 1968. DDT residues and declining reproduction in the Bermuda petrel. *Science,* 159: 979-981.

Pollution—Is There a Solution?

Six papers, three of which appear in this issue and three in the November issue of *BioScience*, were presented in February 1969 at a conference sponsored by the State University of New York and hosted by the Biology Department of State University College at Oneonta. The title of the meeting "Pollution, Public Enemy Number 1" attracted a diverse audience including educators, students, and laymen. There was considerable interest in the sessions as evidenced by an excellent attendance throughout the 2-day meeting and numerous penetrating questions during the discussion periods.

The subject of pollution increasingly has occupied space in national periodicals of both scientific and popular appeal. When the conference was organized originally, an effort was made to have the speakers address themselves to specific areas of pollution with special attention to be given to solutions of the problems. It is evident from these papers appearing in *BioScience* that there are no simple solutions, although each author implies some corrective measures necessary even if they are not specifically mentioned.

As the organizers of this meeting, we feel it is necessary to point out other aspects of the pollution problem which have occurred to us as a result of this conference.

For the first time in man's history he has reached a stage where he actually has the potential to destroy the earth's biosystem either intentionally or accidentally. Yet, generally speaking, the layman refuses even to consider the possibility of disaster. His standard reply is, "Science and technology will resolve our problems, they always have." From the papers contained in this conference and the discussions that followed, we confirmed our personal beliefs that increasingly more scientists do not share this view.

In our opinion, major environmental pollution problems are not going to be resolved simply by scientific and technological advances. The technological means of pollution abatement have been available for decades, but there has been no public insistence on their use. We have the knowledge and techniques to regulate population growth, but any suggestions to require their use is looked upon unfavorably.

What is needed now is an intensive educational campaign designed to increase public awareness of the magnitude of the problem and of the need to make use of available control measures. In this respect the public must share the cost of pollution abatement including costs of research and the implementation of control measures. Further, each individual has the responsibility to consider what *he* can do to relieve the pollution problem including such things as boycotting industrial products from companies which are known to contribute flagrantly to pollution, encouraging political representatives to enact effective pollution control legislation, and personally involving himself in local community environmental quality programs.

We in the educational and scientific community have a special responsibility to become involved in solving these problems. We are perhaps most aware of the gravity of the present environmental situation and thus have the greatest responsibility to assume leadership in the battle for environmental quality. This we can do individually by the organization of and active participation in environmental quality programs in our communities and regions. Hopefully, scientists and educators will meet this challenge *now*, for the deterioration of our environment is increasing at an alarming rate and it is evident that the trend will be reversed only when action begins at the individual and community level.

JOHN G. NEW AND J. GARY HOLWAY
State University College
Oneonta, New York
(BioScience *19*, no. 10, p. 888)

Monitoring Pesticide Pollution

Philip A. Butler

The use of several broad spectrum toxicants presents contamination problems in the environment which may persist for months and even years. Suitable organisms (e.g., the eastern oyster, Pacific oyster, and the soft clam) can be used to give reliable data on the presence of organochloride pesticide residues in estuarine environments. (BioScience 19, no. 10, p. 889-891)

The widespread use of synthetic organochloride pesticides, reports of their persistence in the environment, and the repeated demonstrations of their toxicity to nontarget fauna alerted marine biologists to their potentially disastrous effects in the estuarine environment. Soon after World War II, detailed studies of DDT applications to salt marshes for mosquito control showed increased mortality of fish and shellfish. Such acute effects were readily identified, and measures could be initiated to prevent or at least restrict such applications.

But biologists became even more concerned with the probability that continued terrestrial applications of persistent pesticides would result in their being carried in surface water, adsorbed on silt and debris, through river basins and eventually into estuaries. Here, their chronic presence at subacute levels might cause irreversible changes before their presence was apparent.

The estuary is an extraordinarily important environment to a wide array of fish, shellfish, and other elements of the biota that are important for commercial or esthetic reasons.

Permanent residents of estuaries, such as the oyster, are accustomed to widely fluctuating levels of various environmental properties and are relatively tolerant to unusual changes. Oysters can close their valves and "withdraw" from the environment, when, for example, unacceptable amounts of fresh water or silt are temporarily present. Some animals, however, including many kinds of crabs, shrimp, and fish, use the estuary only as a nursery area or as a part of their migration pathway, and are physiologically adjusted to the estuary for only a particular segment of their life span. As a consequence, they are especially susceptible to drastic environmental changes. Low levels of pollutants might interfere with olfaction, for example, and prevent salmon from finding their "home" stream; and chemicals that changed the osmoregulating ability of crustaceans could prevent shrimp and crabs from migrating to their brackish-water growing areas.

The many possibilities for subtle harmful changes resulting from the accidental or intentional transport of pesticides into estuaries prompted the Bureau of Commercial Fisheries to initiate a program in 1958 to assess the extent of the problem.

The program had two major objectives: to determine the acute and chronic toxicity of the commonly used pesticides to representative estuarine animals under controlled test conditions; and to monitor the seasonal levels of polychlorinated pesticide pollution in the nation's estuaries where production of living marine resources is commercially important. This report describes the development of the monitoring segment of the program and summarizes regional trends in pesticide pollution levels as revealed by 3 years of data collection.

Development of Program

It seemed likely that pesticide pollution in estuaries would be not only intermittent but also at such low concentrations in the water mass that automated sensing devices would be generally unsatisfactory. Consequently, the first requirement for successful monitoring was the selection of a suitable bioassay technique.

The Bureau of Commercial Fisheries Biological Laboratory at Gulf Breeze, Florida, had been engaged for more than 20 years in studies on the ecology and management of the eastern oyster, *Crassostrea virginica*. Much was known about its physiology, and it appeared to be a desirable animal for preliminary tests.

Typically, a mature oyster is feeding 90% of the time and transports about 16 liters of water an hour through its gill system to extract the planktonic food. Its ability and tendency to pick up and store metal ions, present in only trace amounts in the surrounding water, are well known. The oyster is physiologically active the entire year throughout much of its extensive geographical range. Most important, it is sedentary and easily handled.

Exploratory experiments with oysters consisted in recording on a kymograph their shell movements and rates of water transport ("pumping") before and during exposure to controlled amounts of pesticides in aquariums with flowing seawater (Butler et al., 1960). Under normal conditions, the oyster closes its valves briefly up to 10 times an hour to expel accumulated pseudofeces. These experimental

---

The author is with the U.S. Fish and Wildlife Service, Gulf Breeze, Fla.

oysters showed intense physiological irritation after 10 hours exposure to 1 ppm of dieldrin. The valves were almost continually opening and closing. It was obvious that the feeding process could not be normal. The oysters pumped water and fed, however, when the concentration of dieldrin was lowered to 0.1 ppm.

After 2 weeks of continuous exposure to 0.1 ppm of dieldrin, however, the experimental oysters were only half as active as controls. The oysters quickly returned to their normal level of activity when the addition of dieldrin was stopped. This inhibition of the pumping rate in oysters exposed to environmental pollutants can be measured objectively but the technique is tedious and time-consuming.

The interference with normal activity suggested that oyster growth, or rate of shell disposition, would decrease in polluted water. To test this hypothesis, shell deposition in young oysters was measured before and after exposure to known concentrations of pesticides. Linear growth of the oyster does not proceed uniformly. An initial increase in the peripheral deposition of new shell is followed by a period during which this thin shell is strengthened by internal deposits. During this second stage, there may be no measurable increase in length. To encourage the immediate deposition of new shell, the edges of the valves are ground until all new and thin shell is removed. The oyster then occupies all of the shell cavity and has no alternative, in growing, except to deposit new shell on the periphery (Butler, 1965). Linear shell deposition amounts to about 0.5 mm per day for at least 5 days under average conditions in the laboratory, when the oyster has a continuous supply of unfiltered seawater.

In the presence of a toxicant, the oyster deposits smaller and smaller amounts of new shell as the concentration of the toxicant is increased. Thus, it is possible to determine the concentrations of a particular pesticide that causes a 50% decrease in shell growth as compared to control oysters under otherwise similar conditions and to relate the toxicity of one pesticide to that of any other pesticide tested under similar conditions.

In parallel experiments, we found that oysters remove and store chlorinated hydrocarbon pesticides present in the surrounding water at concentrations as low as 0.1 part per billion. Oysters continue to build up such residues in their tissues at uniform rates as long as the toxicant is present. This biological magnification may produce DDT residues, for example, 70,000 times as high as the DDT concentration in the test water supply. An important consideration in a monitor program is the fact that the oyster flushes these residues out of its tissues at a uniform rate when the water supply is no longer contaminated. By sampling an oyster population regularly, it is possible to determine when the water supply becomes contaminated and when the contamination stops. It is possible to get some idea of the magnitude of the pollution load in the estuary by extrapolation from laboratory experiments.

Since oysters are not universally present in estuaries of interest to commercial fisheries, tests were undertaken to determine the relative efficiency of other common mollusks in storing pesticide residues. In general, the eastern oyster, the Pacific oyster (*Crassostrea gigas*), the soft clam (*Mya arenaria*), and any of several mussels are equally suitable. The hard clam (*Mercenaria mercenaria*) is the least efficient at storing pesticides of any mollusks evaluated (Butler, 1966).

With the problem of a suitable bioassay animal solved, we were able to consider the other prerequisites of a successful monitor program. The sampling had to have continuity and be done at least monthly. The assistance and cooperation of many people were required because of the length of the coastlines to be covered. To accomplish analytical uniformity, which was essential, it seemed desirable to have all analyses made by the same laboratory, although this arrangement greatly increased the difficulties in handling the samples.

The program that finally evolved was made possible by entering into formal and informal research contracts with state conservation agencies and university and federal marine laboratories in 15 coastal states. We developed a technique for homogenizing the samples with a desiccant consisting of sodium sulfate plus 10% by weight of Quso, a micro-fine, precipitated silica. This desiccant prevents spoilage of the sample and degradation of pesticide residues for at least 30 days without refrigeration. As a result, samples could be sent by ordinary mail to the Gulf Breeze laboratory for analysis. We analyze samples by gas liquid chromatography with electron capture using two different columns, so that 10 of the most commonly used organochloride pesticides can be quantified at levels above 10 ppb.

In the nearly 4 years the program has been in effect, about 170 permanent stations have been established on the Atlantic, Gulf, and Pacific coasts, and more than 5000 samples have been analyzed.

Summary of Findings

Data collected during the first 3 years of this project do not indicate any consistent upward or downward trends in estuarine pesticide pollution. Distinct seasonal and geographic differences in pollution levels are apparent, however, as well as regional differences in sources and kinds of pesticide pollution. Although each sample is screened for 10 or more pesticides, DDT (including its metabolites) is the only one commonly present. Dieldrin is next in frequency of occurrence, followed by endrin, toxaphene, and mirex.

In the estuaries of Washington, less than 3% of the samples have been contaminated and the DDT residues have always been less than 0.05 ppm. On the Atlantic Coast, oysters from Maine estuaries are the least contaminated; about 10% of the samples had residues of DDT, all at levels below 0.05 ppm. Oyster samples from the other states monitored[1] contained DDT residues more often than not. In some areas that are intensively farmed, oyster samples always contain DDT residues, but the amounts of DDT plus metabolites are usually less than 0.5 ppm. Only rarely has the residue exceeded 1.0 ppm. The highest residue of DDT observed in oysters, 5.4 ppm, was the result of a single incident of gross pollution.

It should be pointed out that these DDT residues are not of sufficient magnitude to constitute a human health problem. Their presence indicates, however, the ubiquity of DDT in the estuarine food web.

---

[1] Alabama, California, Delaware, Florida, Georgia, Maryland, Mississippi, New Jersey, New York, North Carolina, South Carolina, Texas, Virginia.

Analyses of residue data from individual river basins that drain predominantly agricultural lands show characteristic seasonal trends. Typically, there is a pronounced peak in the late spring and a lesser increase in the fall. The height of these peaks may vary tenfold from one basin to another, depending on the farming practices. In one basin on the Texas coast, three seasonal peaks in DDT residue levels reflect the intensive truck farming in an area where the mild climate permits three harvests each year.

In river basins receiving significant amounts of municipal wastes as well as agricultural runoff, the seasonal residue pattern is similar but the magnitude of the residues is proportionately higher. By contrast, in estuaries that receive industrial wastes containing pesticides, seasonal residue patterns are erratic and occasionally indicate single, massive injections of a pollutant into the drainage system such as would happen when waste-control systems are breached. In two areas, the monitor data demonstrated the industrial discharge of pesticide wastes whose presence had been unsuspected by the state agencies responsible for clean water programs.

Several times it has been possible to identify specific sources of pesticide pollution by placing trays of oysters at intervals in the drainage system when natural beds of oysters did not occur. Near and above the river mouth, where the water becomes too fresh for oysters, we have substituted the brackish water clam (*Rangia cuneata*) and the freshwater Asiatic clam (*Corbicula fluminea*) as bioassay animals.

In one locality where a specific pollution source was suspected, trays of oysters were located at about 2-mile intervals upstream toward an industrial complex. Pesticide residues in oysters analyzed during the succeeding 10 months at the several stations, listed in order of their nearness to the pollution source, averaged as follows: 25.0, 17.0, 5.0, 3.0, 0.05, and 0.02 ppm. Analysis of the residue data at any one of these stations demonstrated the intermittency and variation in the amount of waste discharged.

Programs for control of noxious insects involving direct application of insecticides to a salt marsh, constitute still another important source of estuarine pollution. Although the use of the persistent pesticides for this purpose is declining with the increased knowledge of the harmful side effects, DDT is still popular because it is cheap and effective. The 1-pound-per-acre, or less, application rates seem disarmingly small, yet the persistence of this chemical is such that the accumulation of residues in the fauna may reach disastrous levels.

In one bay in New York, for example, DDT residues in soft clam populations are higher by an order of magnitude than at any of 15 other stations routinely sampled on Long Island. This unusually high residue is not caused by the use of more insecticide there for the control of mosquitos, but is a result of the bay being isolated from the sea for much of the year. Since bay water is not diluted by tidal action, the residues accumulate in the environment where they are adsorbed on silt and plankton that the clams ingest as they feed.

A different situation exists in an estuary of northwest Florida, where the stable fly (*Stomoxys calcitrans*) is an obnoxious pest during the summer tourist period. This fly lays its eggs in windrows of seaweed along the beaches, where the larvae then develop. Present control methods consist of spraying the windrows with a DDT formulation three or four times during the early part of the summer. The total amount of DDT used in spraying the approximately 100 miles of beach is relatively small. We calculated that if all of this DDT were evenly dispersed in the estuary at a single time the concentration would be less than 0.001 ppm. Chemical analyses of DDT residues in the plankton following this spray program, however, showed average accumulations of about 0.07 ppm. The DDT is washed into the estuary by tidal action in a relatively short time and, since there is only one tide a day in this area, the flushing rate is relatively slow. As a result, the DDT may persist adsorbed on silt or plankton for a matter of weeks. Plankton makes up the food supply of various small fish including the pinfish (*Lagodon rhomboides*). Analysis of random samples of this fish showed pesticide residues in the range of 0.1 to 0.5 ppm. The level of these residues increased the longer the fish were in the estuary, i.e., the older they were. In turn, pinfish are fed upon by various water birds, including the loon (*Gavia immen*).

One analysis made of the liver of a loon shot in the vicinity showed that it contained about 180 ppm of DDT. Mullet (*Mugil cephalus*) are herbivorous fish that feed largely on plankton and consequently build up relatively large residues of DDT. In one series of analyses, for example, mullet gonads contained 3 to 10 ppm of DDT. At the top of this food chain is the bottlenose dolphin (*Tursiops truncatus*) which feeds extensively on mullet. Blubber samples from a small series of porpoise found dead in this area contained DDT residues ranging up to 800 ppm.

Evidence is substantial in the freshwater habitat that the accumulation of about 4 ppm of DDT in fish eggs may cause complete mortality of the brood. Apparently a similar situation exists in one estuary on the Texas coast. Levels of DDT residues in the ovary of the speckled seatrout (*Cynoscion nebulosus*) were as high as 8 ppm in the prespawning period in 1968, and there was no evidence of successful spawning later in the year. In a second estuary 100 miles away, ovarian DDT residues in seatrout were about 0.2 ppm and a normal number of young of the year were observed.

These data demonstrate the importance of trophic magnification of persistent pesticides in the food web and indicate that even relatively small applications of pesticides in pest control programs can be magnified astoundingly at higher trophic levels. Pesticides are, of course, not all bad. Many are essential to the production and protection of foodstuffs and other agricultural products, in the management of our natural resources, and for the amelioration of our daily existence.

Our difficulties arise from the use of broad spectrum toxicants that may contaminate the environment for months and years. These persistent pesticide chemicals, resistant to decay, may be carried by surface drainage waters to areas far from the point of application. Through the process of co-distillation, they pass into the atmosphere or are airborne on dust particles and carried by winds to every part of the globe, to contaminate even the polar regions and the mid-ocean environment. The use of such pesticides must be limited to those purposes for which there is no substitute. They must not be used on the basis of economy and expediency.

Theoretically, all pesticides are used on a cost-benefit ratio. It appears, however, that even after a quarter century of use and research on DDT, we still do not know the true extent of the damage it has inflicted on the environment. It is essential that laboratory and field tests to evaluate synthetic pesticides be so thorough that we can be assured they will have no irreversible effect on the environment.

References

Butler, Philip A. 1965. Reaction of some estuarine mollusks to environmental factors. In: Biological problems in water pollution, third seminar 1962, p. 92-104. U.S.P.H.S. Publication No. 999-WP-25.

──── 1966. Pesticides in the marine environment. *J. App. Ecol.*, 3 (Suppl): 253-259.

Butler, Philip A., Alfred J. Wilson, Jr., and Alan J. Rick. 1960. Effect of pesticides on oysters. *Proc. Nat. Shellfish Assoc.* 51: 23-32.

Interactions

Mitchell R. Zavon, M.D.

The interactions of those chemicals which are classified as environmental pollutants are many and varied and our knowledge of these interactions is only beginning. Evaluation of the impact of pollutants on the human organism awaits the results of much more research. The interactions of pollutants having been shown to be additive, synergistic, or to cancel each other, it is impossible to predict overall effects with any degree of certainty. (BioScience 19, no. 10, p. 892-895)

Introduction

Dramatic increases in our understanding of biological phenomena which have occurred during the past two decades have made biology the frontier of science. An upsurge of new knowledge of the type we have witnessed has resulted in part in speculation about phenomena—speculation being a prelude to theorization and theorization being the mother of experiment. This ferment of activity and of speculative thought is a healthy sign but one which must be placed in its proper context. Speculation is a long, long step from proof. Theory is not proof. What I will emphasize in this paper is speculation and theorization, but the speculation and theory will be so labeled. The present mode of failing to differentiate between what is speculative and what has been proved is not conducive to the development of the biological sciences. Nowhere is this fault more in evidence than in the area of what is now referred to as "Environmental Health" and "Pollution."

In a conference on pollution we are faced with the necessity for defining our terms before we can deal with the words effectively. Many of the phenomena with which we are dealing are too new or too newly recognized to be generally appreciated by the scientific community. For example, when we speak of air pollution,

Dr. Zavon is with the Kettering Laboratories, University of Cincinnati, Cincinnati, Ohio.

it is meaningless to do so without defining the source of the pollutants and the nature of the pollutants because of the comparatively wide variation. Sulfur dioxide may be of prime concern in one situation and an airborne particulate of natural origin in another. The catalog of actual or potential air pollutants is almost limitless as is true of water pollutants. Any attempt to catalog and discuss the possible interactions of these pollutants would assume encyclopedic dimensions in short order. For this reason alone, if not for my lack of pretense to such encyclopedic knowledge, I will have to resort to a few examples in each situation.

Our present discussions of pollution are too limited in scope and unsophisticated in content. When we discuss the actual or possible effects of air pollutants, we are usually content to talk of the effect of an air pollutant, carbon monoxide or sulfur dioxide, and rarely consider additive effect or the effect of air pollutant and an infectious agent (Anon., 1968). Still less do we consider the effects of noise alone or of noise on the body's response to chemical agents. The possible interactions of foods, food additives, drugs, pesticides, water pollutants and air pollutants (whether naturally occurring or of man-made origin), and electromagnetic energy are innumerable. I do not believe that at this time we can even begin to more than scratch the surface toward understanding the interactions which undoubtedly occur. But I would hasten to add that man appears to have survived as a species through millennia in which he had less understanding than we have now; and I'm rather optimistic that if we do not blow ourselves up, like children playing with gunpowder, we will grow to maturity without undue impairment.

There are some interactions of which we are aware and about which we can speak with some reliability. Perhaps we can begin our discussion with food.

Food

All of us eat to remain alive, although there is some variation in our daily food intake. Millennia of trial and error have helped to teach man what he can eat and what is not safe to eat. But our wisdom is limited. Many of our foods contain naturally occurring materials which are harmful to man or may be harmful under certain circumstances (Food Protection Committee, 1966).

An active goitrogen, 1-5-vinyl-2-thiooxazolidone, is found in the seeds of Brassicae and in the roots of turnip and rutabaga. Cabbage grown on media low in sulfate may not display the same goiterogenic activity, while that grown on media high in sulfate will have higher concentrations of organic and inorganic sulfur and be strongly goiterogenic (Astwood et al., 1949; Sedlak, 1961). These goiterogenic chemicals may be transferred to man in the milk of cows consuming cruciferous plants (Clements and Wishart, 1956). Other chemicals such as thiocyanate and arsenic may contribute to enlargement of the thyroid or may interfere with drugs taken for their therapeutic action on the thyroid.

It is interesting to take this one example and speculate as to the implications as we explore the man-made pollution problem. If sulfate-containing fertilizers are used to grow cruciferae, it is possible that their goiterogenic activity would be enhanced. If, however, the consumer lived near a factory excreting traces of iodine in its air effluent, the iodine might well block the goiterogenic action of the cruciferae. This very simple example is not at all simple. Iodine can block the goiterogenic actions but itself can cause goiter, and the mechanisms involved are not yet completely understood.

Much of what we "know" about the biological effects of food constituents is based on the response of the rat. Aflatoxin, a mixture of several compounds characterized as complex difurano-coumarins (Fig. 1), is carcinogenic for the rat and has been found to induce hepatomas in rainbow trout. Diets containing 0.5 ppm of aflatoxin throughout a lifetime cause 100% incidence of liver tumors in male rats. The incidence in the female is somewhat less. At 0.08 ppm in the diet, aflatoxin appears to cause hepatomas in the livers of rainbow trout. The incidence of tumor and the dietary level required will vary, depending on the strain of rat and on the mixture of aflatoxin, a toxin formed by a strain of *Aspergillus flavus*. The concentrations mentioned are far from the most potent found but are sufficient to illustrate that this naturally occurring material is one of the most specific carcinogens known as well as one of the most toxic by other criteria. A single oral dose of aflatoxin may produce liver cancer in a rat surviving for 2 years. Thus far monkeys have not reacted as does the rat and trout, and we do not know man's response. There is, however, sufficient justification for eliminating this fungal metabolite from man's food (Sargeant et al., 1963; Nesbitt et al., 1962; Ashley et al., 1964; Lancaster et al., 1961; and Butler and Barnes, 1968).

These examples of problems with "naturally" occurring toxicants lead us into the problem of man-made substances. The emphasis, until comparatively recently, on the toxicology of man-made materials has lead to the neglect of the study of naturally occurring materials. If we do not understand the interactions of the basic dietary constituents, how can we draw valid conclusions about the effects of added constituents?

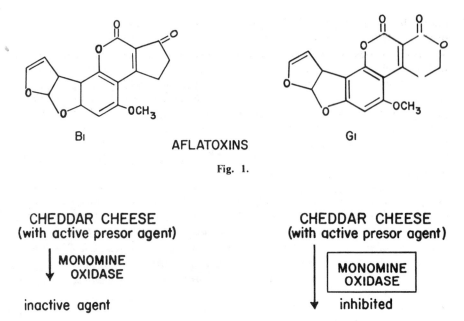

Fig. 1.

Fig. 2.

Food Additives

Chemicals not native to the food stuff in question may be deliberately added to extend the period during which the food may be eaten, make it more enticing to the consumer, or impart properties which make the food stuff easier to handle during processing. None of these reasons are necessarily amoral, as some would have us believe, or without any possible harm, as others would like to believe. Naturally occurring compounds, as has already been described, may cause problems, so that what we are discussing is not necessarily unique to man's own actions.

Food additives may cause inadvertent difficulties far removed from the actual consumption of food. For example, Erythrosin BS, an iodine substituted fluorescein, is used in both foods and pharmaceuticals. Its ingestion may increase the concentration of protein-bound iodine in blood serum to what are usually recognized as pathological levels. The protein-bound iodine concentration as a measure of thyroid function could thus be made useless (Keiding, 1964).

The story of butylated hydroxytoluene (B.H.T.) is worth noting as we speculate on interrelationships, and emphasizes once again the need for good science mixed with a generous dose of humility. B.H.T. is used as an antioxidant in oils, fats, and fatty foods and may be transferred as an incidental additive from those medicinals and plastics in which it is used as a stabilizer. Early study of the metabolism of this compound failed to recognize the large amount excreted in the feces of the experimental animal as well as some metabolic degradation products that were soon revealed. Different results in different laboratories lead simultaneously to an attempt to resolve the true state of affairs and at the same time to newspaper accounts and politically inspired investigations of hazard. Eventually, the facts were carefully and rationally revealed but not without grave damage to some scientific reputations. B.H.T. does not appear to be particularly toxic and is not of great concern as a chronic "pollutant" unless one fails to review all of the data critically (Anon., 1965).

The interactions between food additives and between food additives and other substances, naturally occurring or man-made, is relatively unexplored territory. We do know that what may be called the "cheddar cheese syndrome" is caused by eating cheddar cheese within a couple hours after taking a monoamine-oxidase inhibitor. The monoamine-oxidase inhibitors are commonly prescribed medications for control of hypertension. A pressor substance normally destroyed by the monoamine-oxidase (catechol amine) is not destroyed and causes a hypertensive crisis. I doubt that this could have been picked up in animal experiments except by merest chance, yet we must contend with these unexpected occurrences as we work with more specific food additives and more specific drugs. In this instance our equation was as shown in Figure 2 (Blackwell, 1963). We have a naturally occurring food stuff causing an unnatural physiological response because of the intervention of a therapeutic agent prescribed for good and sufficient reason.

We can explore this path a bit further in connection with the use of antioxidants as possible therapeutic agents for the prolongation of life. Mice receiving 0.5% B.H.T. in their diet achieved a 30-40% increase in mean longevity and a smaller gain in specific age (Harman, 1968). If this line of research is pursued successfully, many questions about interactions with other drugs will eventually have to be answered. Adverse reactions are possible but the questions must be answered by experiment, not by speculation.

Pesticides

The widespread use of pesticides since the close of World War II has resulted in insecticide residues in the fat of humans (Zavon et al., 1965). The concentrations of the chlorinated hydrocarbon insecticides found in body fat are easily detectable with the present analytical methodology and are summarized in Table 1 (Zavon et al., 1965). At present, we have no data which indicate that these chlorinated hydrocarbon residues in human fat are, in and of themselves, harmful to the human organism. We do know that they may cause an increase in the production of liver microsomal enzymes which accelerates the metabolism of a wide variety of exogenous agents. For example, rats have been shown to metabolize hexobarbital more rapidly than normal for 65 to 90 days after a single intraperitoneal injection of DDT (Ghazal et al., 1964; Conney, 1967). On the other hand, it is also possible for these same enzymes to cause a substance such as acetylaminofluorene (AAF) to become a carcinogen as a result of metabolic degradation (McLean, 1965).

Organophosphorus insecticides are generally shorter lived than the members of the chlorinated hydrocarbon group. In some instances, as with parathion, they must be metabolically converted to paraoxon in order to become active in the organism. The action, inhibition of cholinesterase, is well known and well described (Riker, 1953). With continued ingestion of low levels of organophosphorus compounds in the diet, it has been shown experimentally that resistance to the action of the compound appears to develop due to the refractoriness of the cholinergic receptors (Dubois, 1965). In considering the interaction of residues of pesticides or food additives with the drugs which may be supplied for good therapeutic reasons the possibility must be considered of interference with end organ receptivity by the chronically applied molecule.

The experimental production of an effect that is more than additive has been described when two organophosphorus compounds are administered simultaneously (Frawley et al., 1957). When malathion degradation is interfered with, it becomes as toxic as parathion (Frawley et al., 1957). In actual usage it is doubtful that this observation has proved to be of real significance. Nevertheless, the existence of this phenomenon must give us pause, for it is impossible to predict the many possible additive actions which are possible. Nor is it possible to predict the individual variability which may produce the exceptional response (Williams, 1956). We must accept the existence of a mode, recognize that variations exist around the mode, and attempt by a series of approximations to refine our predictive abilities. The possibility of extreme variation will always exist and must be looked for, but the existence of variation cannot be used as an excuse for a do-nothing, change-nothing policy.

As we look for interactions, we will find apparent contradictions which have good explanations as we learn more about the biological mechanisms involved. Potentiation, a greater than additive effect, may occur when one organophosphate interferes with the degradation of another compound of the same class. Other phenomena occur, the acute toxicity of an organophosphorus insecticide, dimethoate [o, o-dimethyl S-(N-methyl carboxymethyl) phosphorodithioate], to mice was increased from an LD_{50} of 198 mg/kg to 58.5 mg/kg by pretreatment of mice for 3 days with 75 mg/kg of sodium phenobarbital per day. On the other hand, the *same* pretreatment *reduced* the toxicity of other members of the same class of compounds by approximately one-half. Phosphamidon, Bidrin ®, and their N-dealkylated derivatives all had their toxicity reduced by approximately one-half.

The phosphorothioate and phosphorodithioate insecticides must be activated to become strong cholinesterase inhibitors. This involves the oxidation of the P = S to

TABLE 1. Some chlorinated hydrocarbon pesticides in perirenal fat of persons in four U.S. cities (ppm)

Pesticide Chemical	Total (No. = 64)		Men (No. = 41)		Women (No. = 23)	
	Range	Mean	Range	Mean	Range	Mean
Total chlorinated hydrocarbons analyzed [a]	2.38 to 22.36	7.55	....	....	....	....
Dieldrin	0.07 to 2.82	0.31	0.07 to 2.82	0.31	0.1 to 0.74	0.24
DDE	0.86 to 14.7	4.63	1.10 to 14.7	5.39	.88 to 8.2	3.61
o,p'-DDT	0.01 to 0.76	0.16	0.01 to 0.76	0.21	.01 to 0.20	0.11
p,p'-DDT	0.36 to 8.4	2.35	0.36 to 8.40	2.52	.68 to 4.20	2.18
Heptachlor epoxide	0.03 to 0.61	0.10	0.03 to 0.61	0.11	0.01 to 0.18	0.09

[a] Other chlorinated hydrocarbons may have been present but were not detected by the analytical procedure used, which was gas liquid chromatography with electron capture.

P = O by the microsomal enzymes of the liver. The pretreatment with phenobarbital decreases the toxicity of these organophosphorus insecticides by stimulating the microsomal enzymes (Dubois and Kinoshita, 1965; Menzer and Best, 1968).

Another instance of interaction is the effect of hexachlorophene bathing of newborn infants on reducing infection by *Staphylococcus aureus*. At the same time there has, as a result, been a marked increase in the incidence of gram-negative bacterial infections in these same newborn (Light et al., 1968; Forfar et al., 1968). Control of one organism may, and has, resulted in overwhelming infection by a competitor. In this instance the possibility of direct stimulation of gram-negative organisms by hexachlorophene has not been ruled out.

Drugs

Reference has already been made to drug, food, and food additive interactions. There can be little doubt that more such interactions occur than we can presently catalog. As our drugs gain in specificity and thereby gain in potency, we are more likely to encounter and recognize such interactions.

Recently, interest has been aroused in the possible effects of drugs and other chemicals in the production of genetic abnormalities. Under experimental conditions d-Lysergic acid diethylamide may be mutagenic but whether, under conditions of use, the same can occur in man remains to be determined (Fitzgerald and Dobson, 1968; Cohen and Marinello, 1967; Light et al., 1968).

Air Pollution and Water Pollution

Some of the problems already discussed are also factors in evaluating the effects of air and water pollution. The vast array of industrial chemicals that can be detected in our air and water on one occasion or another are beyond our scope today. The more commonly looked for air pollutants: SO_2, oxides of nitrogen, carbon monoxide, peracetyl oxide, ozone, and lead are all well described. Coffin (Anon., 1968) has described an interaction between ozone and increased mortality from infection in mice. Apparently, at least in the mouse, there is a decrease in the activity of pulmonary alveolar macrophages when the animal is exposed to 0.08 ppm of ozone for 3 hr, NO_2 at 3.5 ppm for 2 hr, and acute exhaust fumes for 4 hr at a total oxidant level of 0.15 ppm.

We must consider the total chemical assault on the organism, irrespective of the route of exposure, if we are to properly evaluate the impact of the pollutant. The uranium miner is exposed to radon and radon daughters, all radioactive, in the mine. In addition, he is likely to be drinking water that contains a greater than usual amount of radium and may well live in an area with somewhat increased radioactive air content because of dust-containing radioactive materials. Each source of radiation exposure must be considered in evaluating the individual's total exposure. This is not an interaction between chemical insults but a summation of sources of exposure.

This last example of what could be a complicated air pollution problem need not be unique because the same situation may prevail in relation to other water pollutants and air pollutants. Fluorides may be present in the stack effluent from a phosphate plant and also be present in the water effluent of an aluminum plant. Numerous other examples might be cited but this should suffice.

Noise

Noise pollution is a recently popularized phrase. Although we have referred to radiation, the effects of the other parts of the electromagnetic spectrum may well give room for speculation. How does a high noise level affect our response to a drug or to an air pollutant? Do we respond differently to 50 ppm of carbon monoxide in the cab of a noisy truck as compared to a quiet automobile? These interactions or possible interactions will have to be explored. We must learn to accept the existence of knowledge gaps until those gaps can be closed. We need not become frantic or hysterical as we try to fill in the many gaps in our knowledge.

Conclusion

Each advance results in newly visible gaps in our knowledge, and this is the excitement inherent in extending our knowledge, not a cause for cries of alarm.

Adlai Stevenson, shortly before his death, said, "We travel together, passengers on a little space ship, dependent on its vulnerable supplies of air and soil . . . preserved from annihilation only by the care, the work, and . . . the love we give our fragile craft." Our fragile craft must be protected, but it must also recognize that only a radical would believe that change can be stopped.

References

Anon. 1965. The eventful history of B.H.T. *Lancet*, November 20, **1965**: 1056.

Anon. 1968. Pollutants make infection deadlier. *J. Amer. Med. Assoc.*, **206** (7): 1432.

Ashley, L. M., J. E. Halver, and G. N. Wogan. 1964. Hepatoma and aflatoxicosis in trout. *Fed. Proc.*, **23**: 105.

Astwood, E. B., M. A. Greer, and M. G. Ettlinger. 1949. 1-5-vinyl-2-thiooxazolidone on antithyroid compound from yellow turnip and from *Brassica* seeds. *J. Biol. Chem.*, **181**: 121.

Blackwell, B. 1963. Hypertensive crisis due to monoamine-oxidase inhibitors. *Lancet*, October 26, **1963**: 849.

Butler, W. H. and J. M. Barnes. 1968. Carcinogenic action of groundnut meal containing aflatoxin in rats. *Food Cosmet. Toxicol.*, **1968** (6): 135-141.

Clements, F. W., and J. W. Wishart. 1956. A thyroid blocking agent in the etiology of endemic goiter. *Metabolism*, **1956** (5): 623.

Cohen, M. M., and M. J. Marinello. 1967. Chromosomal damage in human leukocytes induced by lysergic acid diethylamide. *Science*, **155**: 1417.

Conney, A. H. 1967. Pharmacological implications of microsomal enzyme induction. *Pharmacol. Rev.*, **19** (3): 317.

Dubois, K. P. 1965. Low level organophosphate residues in the diet. *Arch. Environ. Health*. **10**: (6): 837.

Dubois, K. P., and F. Kinoshita. 1965. Modification of the anticholinesterase action of o, o-diethyl o-(4-methylthio-m-tolyl) phosphorothioate (DMP) by drugs affecting hepatic microsomal enzymes. *Arch. Int. Pharmacodyn. Ther.*, **156**: 418-431.

Fitzgerald, P. H., and J. R. E. Dobson. 1968. Lysergide and chromosomes. *Lancet*, May 11, **1968**: 1036.

Food Protection Committee. Food and Nutrition Board, National Academy of Sciences, National Research Council. 1966. Toxicants occurring naturally in foods. Publication 1354.

Forfar, J. O., J. C. Gould, and A. F. MacCabe. 1968. Effect of hexachlorophene on incidence of staphylococcal and gram-negative infection in newborn. *Lancet*, July 27, **1968**: 177.

Frawley, J. P., H. N. Fuyat, E. C. Hagan, J. R. Blake, and O. G. Fitzhugh. 1957. Marked potentiation in mammalian toxicity from simultaneous administration of two anticholinesterase compounds. *J. Pharmacol. Exp. Ther.*, **121** (1): 96-106.

Ghazal, A., W. Koransky, J. Portig, H. W. Vohland, and I. Klempaw. 1964. Beschleunigung von Entgiftungsreakitionan durch verschiadene insecticide. *Arch. Exp. Pathol. Pharmakol.*, **249:** 1-10.

Harman, D. 1968. Free radical theory of aging. *Gerontologist,* **6:** 12 and *J. Gerontol.,* **23:** 476.

Keiding, Rud. 1964. Five years grace. *Lancet,* September 12, **1964:** 587.

Lancaster, M. C., F. P. Jenkins, and J. M. Philp. 1961. Toxicity associated with certain samples of groundnuts. *Nature,* **192:** 1095.

Light, I. J., J. M. Sutherland, M. L. Cochran, and J. Sutorius. 1968. Ecologic relation between *Staphylococcus aureus* and *Pseudomonas* in a nursery population. *N. Engl. J. Med.,* **278:** 1243.

McLean, A. E. M. 1965. Pesticides and food additives. *Lancet,* December 18, **1965:** 1295.

Menzer, R. E. and N. H. Best. 1968. Effect of phenobarbital on the toxicity of several organophosphorus insecticides. *Toxicol. Appl. Pharmacol.,* **13:** 37-42.

Nesbitt, B., J. O'Kelly, K. Sargeant, and A. Sheridan. 1962. Toxic metabolites of *Aspergillus flavus. Nature,* **195:** 1062.

Riker, W. F., Jr. 1953. Excitatory and anticurare properties of acetylcholine and related quaternary ammonium compounds at the neuromuscular junction. *Pharmacol. Rev.,* **5:** 1.

Sargeant, K., R. B. A. Carnaghan, and R. Allcroft. 1963. Toxic products in groundnuts. In: *Chemistry and Origin,* Vol. II, p. 53, Chemical Industries, London.

Sedlak, J. 1961. Cultivation of goitrogenous and nongoitrogenous cabbage. *Nature,* **192:** 377.

Williams, R. J. 1956. *Biochemical Individuality.* John Wiley & Sons, Inc., New York.

Zavon, M. R., C. H. Hine, and K. D. Parker. 1965. Chlorinated hydrocarbon insecticides in human body fat in the United States. *J. Amer. Med. Assoc.,* **193:** 837.

Some Effects of Air Pollution on Our Environment

Vincent J. Schaefer

Weather systems are changing as air pollution stabilizes clouds, which, in turn, prevent sunshine from reaching the earth. Although air pollution may or may not be damaging the atmosphere enough to affect the climatic patterns of the world, it will result in serious environmental and ecological problems. (BioScience 19, no. 10, p. 896-897)

The rapid increase in air pollution is a fact that even the casual observer can see—and often smell. In some areas it is so bad that eye-watering conditions from smog are not uncommon. This condition has happened and is actually getting worse despite considerable effort on the part of air pollution control officials, industry, and the enforcement of a number of local laws controlling trash burners, brush burning, and other practices.

Many observers are puzzled about this buildup of pollution especially when they notice that visible plumes from chimneys and smoke stacks are rarely seen except from electric power plants, steel and pulp mills, cement plants, and some chemical plants.

One of the major sources of air pollution consists of invisible plumes of particulates so small, as they emerge from the combustion chamber, chemical reaction, or gaseous vapor source, that they are optically invisible. Such particles have cross sections less than $0.1\ \mu$. One source of such particles is the automobile. When in good operating condition, the effluent from the auto exhaust pipe is quite invisible. However, if one measures the number of particles emitted by an idling automobile, it is the order of one hundred billion (1×10^{11}) particles per second. Another potent source of invisible particles may actually result from an air pollution law which is directed at the control of visible smoke plumes. While this law was designed to force industrial plants to install electrical precipitators, scrubbers, and other smoke control devices, it is possible in some instances to pass the effluent from an industrial process through a hot flame (an afterburner) to vaporize it and thus as with the automobile, the pollution plume becomes invisible. The concentration of tiny particles is so high, however, that agglomeration often occurs and the knowledgeable observer will detect the plume downwind of the offending source. Under such conditions the use of the afterburner is particularly bad since in addition to making the particles much smaller than they would normally be, they then have a longer residence time in the atmosphere because of their smaller size. Also, the afterburner will generate nitrogen oxide, a poisonous gas which also serves as one of the catalysts for particle growth involving unburned gasoline vapor.

Although some persons believe that unless pollution is curbed in the near future we will run out of breathable air, I believe that other problems will confront us before that happens. The human body is a highly resistant mechanism to airborne particles. If this were not the case, I do not see how smokers could live! In the process of smoking, the individual insults his lungs with a concentration of at least ten million smoke particles per cubic centimeter. This is a concentration that is 10 to 100 times greater than is encountered in a very badly polluted urban area like Los Angeles or New York City. While there is increasing evidence that the smoker is, in fact, shortening his life by the act of smoking, there are many contradictory facts about smoking which require more understanding about this complex question.

In considering this problem I have called the cigarette a "synergistic reactor." By this term I mean the following: when a cigarette is smoked, there is a very hot zone at the site of the burning tobacco. When the smoker inhales, this burning zone increases in temperature as more air ventilates and intensifies the burning of the tobacco. If the cigarette is smoked in air which contains pollution particles, many of these particles (with a concentration of 10,000 to 100,000 per cubic centimeter or more) are drawn through the burning zone and vaporized. Thus, besides receiving the products of combustion of the tobacco and paper of the cigarette, an additional load comprised of a wide variety of chemical substances is also taken into the lungs. Through vaporization these chemical substances are now in a highly reactable condition with the lungs virtually serving as a test tube, the concentration of gaseous vapors being so high that many reactions can take place and consequently a host of new chemicals may form. These new precipitates being small are readily adsorbed, dissolved, or precipitated on the moist surface of the lungs.

When one considers the nearly infinite variety of substances which float in the air

---

The author is with the Atmospheric Sciences Research Center, State University of New York at Albany.

of the urban environment, is it any wonder that confused information is an inherent part of the health records of smokers in an urban region?

One of the disturbing aspects of the increase in air pollution over the past decade is that it has apparently increased by nearly an order of magnitude in the area upwind of our cities. This tenfold increase in particulates in areas which previously were characterized as clean "country" air has been measured in northern Arizona near the San Francisco Peaks, in northwestern Wyoming at the Old Faithful area of Yellowstone National Park, and in the Adirondack Mountains of northeastern New York.

When the country air becomes contaminated, then it can no longer dilute the pollution sources to the degree which once was possible.

During the past 3 years we have measured the concentrations of particulates in many parts of the contiguous United States (Schaefer, 1969). In eight transcontinental flights which encompassed most of the major cities of the country and the majority of the clean and polluted regions of the country in between, we have been able to gain a very broad view of the degree to which polluted air covers the country. These findings show the extent to which pollution sources spread their pall over large areas of the continent and clearly shows the fallacy of the "air shed" idea. Unlike water which is primarily controlled by gravity and thus can be related to a particular drainage system often called a water shed, the air and its load of particulates is not controlled by geographical barriers in most instances as it moves rapidly from one region to another, controlled primarily by pressure systems and the weather accompanying them. It is only during periods of clear, quiet weather that local inversions intensify the pollution loads and thus cause local concern as the concentration of particles builds downward from the lid of the temperature inversion and the general public becomes aware of the pollution haze and is sometimes frightened by it.

It is the measurable increase in the continental and global levels of particulates which concerns me at the present time, since I believe certain components of the polluted air may affect us in a more subtle way which may become a more serious problem than the foul smell and eye-watering components we now associate with polluted air.

For a quarter of a century I have been observing, studying, and measuring the characteristics of atmospheric clouds. In the mid-1940's, supercooled clouds were frequently observed in the northeastern United States. During the period 1946-52, we conducted more than 150 flight studies of supercooled clouds in eastern New York, mainly over the Adirondack Mountains. During the past 5 years, supercooled clouds have become relatively rare in this same region while low-level ice crystal clouds (false cirrus) are of common occurrence.

This disappearance of supercooled clouds has been accompanied by the occurrence of a strange kind of precipitation which I have called "misty" rain and "dusty" snow. The mist consists of water droplets of about 0.050-cm diameter, so small that they tend to drift down rather than fall.

The "dusty" snow has a slightly larger cross section, but the droplet from a melted crystal has about the same size as the misty rain. Thus, it is likely that some of the mist originated as snow but melted as it fell into warm air. When this type of snow is falling, only a thin dust-like layer of snow accumulates on the ground.

I believe the origin of both of these forms of precipitation are produced by air pollution. A superabundance of both cloud nuclei and nuclei for ice crystal formation are commonly observed in polluted air. The water vapor in the air collects on such particles, but since there are so many of them (often 10 to 20 times more cloud nuclei and hundreds of times more ice nuclei), the particles are so numerous and so small that they inhibit the precipitation process by stabilizing the clouds. Such stable clouds prevent sunshine from reaching the earth and thus may effectively change some of the dynamics of weather systems. Whether such effects will eventually cause changes in climate can only be determined by much more intensive research.

The type of air pollution which produces a large increase in the concentration of cloud nuclei could be almost any smoke from the burning of organic materials such as garbage, wood, paper, the effluent from pulp mills, a majority of chemical plants, and electric power plants emitting sulfur residues. On the other hand, only a few sources of ice crystals have been identified. A very definite increase in such nuclei may be found in the smoke from steel plants (Chagnon, 1969). By far the greatest and ubiquitous source is the automobile. The most effective source is the auto whose exhaust is quite invisible. I have recently found that an auto equipped with the so-called anti-pollution devices is as effective a source of potential ice nuclei as a car without such a control mechanism. If anything, it appears to produce even more nuclei! The material responsible for the production of ice crystals is the submicroscopic residues produced by the burning of gasoline. This mechanism and the results have been previously described in some detail (Schaefer, 1966; Hogan, 1967; Schaefer, 1968a and b) and will not be repeated here.

If our studies continue to show the increase in occurrence of overseeded clouds, the persistence of clouds for longer periods of time due to their stabilization in areas of polluted air, there is a strong possibility that such conditions will lead to serious environmental and ecological problems resulting from this inadvertent modification of the weather and precipitation.

I hope I am wrong about these mechanisms and their consequences. However, the more data I accumulate and the more observations I make increases the evidence that some major effects in the atmosphere are occurring over hundreds of thousands of square miles. Since the weather systems of our planet are interconnected on a global scale, these effects may lead to an ever-increasing impact on the climatic patterns of the world. While such effects are not necessarily irreversible, it would require major changes in the present trend of our scientific and technological developments to reverse the present situation.

Only an educated and aroused public is likely to demand a change in this deterioration of our environment. There are some hopeful signs that the public is aware of some of these abuses and dangers. Many more efforts like the "Conversation in the Disciplines" in which we have participated are needed. I hope this will happen; otherwise we will encounter a host of serious problems within the next generation.

References

Chagnon, S. A. 1969. La Porte weather anomaly, fact or fiction? *Bull. Amer. Meteorol. Soc.*, **49**: 4.

Hogan, A. 1967. Ice nuclei from direct action of iodine vapor with vapors from leaded gasoline. *Science*, **158**: 800.

Schaefer, V. J. 1966. Ice nuclei from automobile exhaust and iodine vapor. *Science*, **154**: 1555.

———. 1968a. Ice nuclei from auto exhaust and organic vapors. *J. Appl. Meteorol.*, **7**: 113.

———. 1968b. The effect of a trace of iodine on ice nucleation measurements. *J. Rech. Atmos.*, **3**: 181.

———. 1969. The inadvertent modification of the atmosphere by air pollution. *Bull. Amer. Meteorol. Soc.*, **50**: 199.

Technology Assessment

Ann C. Barker

and

Jo Ann Fowler

(BioScience *19,* no. 10, p. 923-924, 934-935)

At the request of Representative Emilio Q. Daddario (D-Conn), chairman of the House Subcommittee on Science, Research, and Development, the Legislative Reference Service of the Library of Congress recently completed a study and issued a report "Technological Information for Congress." The report is actually 14 case histories which were studied in depth to determine the method and the effectiveness of the scientific and technological briefings Congress receives from the scientific community. Below we will discuss one of the 14 cases, "The Insecticide, Fungicide, and Rodenticide Act of 1947,"[1] and a 15th case reported separately, "A Technology Assessment of the Vietnam Defoliant Matter."

In general, the study concludes that some kind of early warning system must be devised in order to set the Congressional wheels in motion as soon as any technology advancement shows its Jekyll-like characteristics. Only by beginning studies much sooner than heretofore can Congress operate on an orderly time schedule with "less reliance on crash decision-making and a reduced frequency of sudden sensational alarms." The authors conclude that unless technical questions are resolved first, political decisions are almost always defective. They also found instances where unsatisfactory decisions resulted from a failure to adhere strictly to the scientific canon which says that "only after rigorous investigation and overwhelming supporting evidence" does hypothesis become fact.

Another obstacle to sound action is the relationship between technical decision-making and sensationalism. Sensationalism, they say, serves a useful purpose—it focuses attention on a particular problem—but it also creates an atmosphere in which emotionalism overrides calm deliberation and tends to create a desire for hasty action. This is also true of flamboyant and sensational personalities as witnesses—they give the issue public visibility but tend to divert attention from the real problem and onto themselves.

"Scientists sometimes disagree as to the facts . . ." and "frequently disagree as to the correct interpretation of the facts," the authors note. This problem is usually solved by further examination of the data by other scientists. Nevertheless, it is very frustrating for the Congress when there are persistent disagreements among the experts. "The ability to analyze scientific evidence and draw valid conclusions is not evenly distributed among scientists," the authors lament. "In short science is no royal road to truth."

The selection of scientific witnesses is another difficulty. One finds the same witnesses appearing again and again before congressional microphones. These are usually persons of recognized eminence and they are particularly useful because, over the years, Congress has come to have confidence in them. Witnesses, however, are seldom impartial. Scientists, like all others, have their own prejudices, be they personal, political, social, or intellectual, and so it is often difficult to strike the right balance in selecting witnesses. It was once possible to classify witnesses along simplistic lines, e.g., adherence to one or another political party, industry versus the academic, technical background versus liberal arts, and so on. The authors suggest that this is now outmoded. Partisan affiliations, they say, are irrelevant to most technical issues, and the frequent shifting of personnel between government, business, and academic institutions together with the consulting roles to both government and industry undertaken by many academicians, all tend to blur the distinction. It is suggested that if it is possible to categorize present-day witnesses some possible categories are "mission-oriented versus discipline-oriented," "specialist versus generalist," "technocrat versus antiscience," and "economic emphasis versus ecological emphasis."

The authors suggest a number of ways in which the Congress might secure faster, more timely, and dependable technical and scientific information in the future. These include additional staff for individual Congressmen or Committees, the development of an entirely new mechanism, or the expansion of the Legislative Reference Service. One of the last two seem to be the most logical, since the present fragmentation will only be compounded by either of the first two. The authors appear to favor this approach, but whatever the ultimate decision they urge that any service established or designated to assess technology and to serve Congress as an early warning system maintain close liaison with the scientific community in colleges and universities, in professional societies, in private industry, and elsewhere. It is also suggested that the professional staff of such a service might contribute papers to the scientific journals and at symposiums and seminars, and that there might be an interchange of personnel between the service and academic research institutions on a temporary basis.

The Pesticide Controversy

The history of the pesticide controversy is too well known to require amplification here,

---

Both reports used may be obtained from the House Committee on Science and Astronautics, 2321 Rayburn House Office Building, Washington, D.C.

[1] The other 13 cases are: AD-X2 (a battery additive); the Point IV program; inclusion of the social sciences in the scope of the National Science Foundation; Project Camelot; the decline and fall of Mohole; the test ban treaty; establishment of the Peace Corps; high-energy physics; the Office of Coal Research; the Salk Vaccine; the Water Pollution Control Act of 1948; Thalidomide; and Congressional decisions on water projects.

but for anyone who may have been stranded on a desert island for the past year or so, stated briefly, it is this. Toward the end of World War II the new science of synthetic organic chemistry made possible the development of pesticides which since that time have been in world-wide use on a huge scale. These new pesticides were cheap, effective, and probably safer than the dangerous arsenicals they replaced, but some species of insects quickly developed resistance to them. The user simply counterattacked by spraying with greater frequency and in greater quantity, or the chemical industry obliged with a more persistent compound.

These pesticides have aided enormously in the production of food and have saved countless lives through the control of disease-carrying insects, but the question now arises as to their long-term effect on the total environment. There was some protest from ecologists from the beginning; but in 1962 the publication of Rachel Carson's book *Silent Spring* focused public attention on the possible hazards in the indiscriminate use of pesticides. By 1967 the foes of pesticides were in full cry and the controversy continues today.

The Insecticide, Fungicide, and Rodenticide Act of 1947 provided for registration of economic poisons, appropriate labeling of containers (including instructions for safe use), and the addition of color to poisonous white powders to distinguish them from harmless ones. In studying the 1947 Act, the Legislative Reference Service hoped to answer such questions as whether the potential threat should have been manifest when the hearings were conducted in 1946-47; whether the Department of Agriculture should have recognized the need for action; and whether the growing importance of preserving the environment should have been recognized, given the state of scientific knowledge at that time.

Hearings began into the pesticide legislation in February 1946 before the House Committee on Agriculture. The bill was described as one to "regulate the marketing of economic poisons and devices, and for other purposes." During four days of hearings at which 16 witnesses were heard, "No controversy of consequence developed . . . " according to the study. Representatives from trade associations, industrial producers, associations of agriculturists, state government, and one professional association (American Association of Economic Entomologists) were heard. The bill was reported favorably but no further action was taken. In 1947 the measure was reintroduced. Hearings this time took only one day and once again no substantial controversy developed. However, one witness in 1947 hinted that insecticides might kill useful as well as pestilent insects, and suggested that manufacturers might be required to issue appropriate warnings. He added that he recognized that this suggestion was premature. His name was William Heckendorn and he was representing The National Council of Farmer Cooperatives. Having told the Committee that he realized that the question of hazard "cannot be taken into consideration in this particular bill because we do not know enough about pesticides yet," he went on:

"You recall a few years ago a group of us came before you, and asked for a special appropriation of $12,500,000 for incentive payments in the production of legume seeds. Since that time, we have set up a program under the Agriculture Research Administration, trying to determine why it is that our yields of legume seeds have dropped so rapidly.

"One of the reasons . . . is the fact that our insecticides will kill everything. They kill both our beneficial insects as well as our harmful insects. And so far our development in insecticides has been to kill; it has not been to try and isolate and use certain insecticides that will protect our beneficial insects.

"I feel the time is coming when we are going to be obliged to give more consideration to the type of insecticides which we use simply because we now find bee keepers are unwilling to place their bees in areas where certain insecticides are being used, simply because their colonies are being killed off." And he concluded, "I may be raising a question here that might create quite a controversy."

Mr. Heckendorn worried prematurely, for 15 years were to elapse before the controversy really got into stride. It is recorded, however, that a series of specific investigations were conducted as early as 1943 and 1944, and in 1945 the Departments of Agriculture and Interior sponsored more elaborate studies into the ecological effects of DDT. Here are some of the comments from a summary report which made its appearance in 1946:

"Spray drifting about 150 feet from the sprayed area to a small gravel-pit pond killed all the golden shiners and pumpkin-seed sunfish in it."

"Within 48 hours after the application of DDT to the final portion of the area on June 1, the bird population (which had been 1.6 pairs to the acre before spraying) was much reduced. On June 18, the area contained only 0.5 birds to the acre."

"Most of the crayfish on the area were readily killed by the DDT solution applied at 0.5 pound to the acre."

The Legislative Reference Service found that students of ecology were expressing concern at least as early as 1946, and with increasing frequency thereafter. In the medical profession, too, the complexity of the effect of poisons and drugs on the human organism was being recognized.

After 1947 the authors find a steadily increasing volume of reports, investigations, and protests concerning the use of pesticides, which reached a peak with the publication of *Silent Spring* in 1962 and continues unabated to this day. Apparently the only evidence which might have drawn attention to the possible danger at the time of the 1946-47 hearings was the summary report on DDT which appeared in 1946, and Mr. Heckendorn's gentle demurer. The authors also tell us that ecologists protested at an early date but they do not say to whom, and one suspects it may have been to each other.

In any event, the House Agriculture Committee favorably reported the measure a few weeks after the 1947 hearings and it passed the House the following month. The Senate dealt even more summarily with the matter, holding no public hearings at all. The bill became Public Law 80-104 on 25 June 1947.

The Vietnam Defoliant Matter

The 15th case history of technology assessment deals with the controversy over the military use of chemical defoliants and herbicides in Vietnam and is distinguished from the other case histories in that the assessment was performed by the scientific community itself. As the report notes, "it does not address itself to the merits of the issue; it does not judge the propriety or impropriety of the military use of herbicides," but rather traces the evolution of dissent on the question; the exchange of views by the American Association for the Advancement of Science within its own governing bodies and with the Department of Defense; DOD's efforts to perform its own assessments; the efforts of various scientists to perform an assessment; and the present status of the defoliant assessment. In releasing the report, Mr. Daddario explained that at his direction the Legislative Reference Service "secured the status of observer to the proceedings in which the AAAS undertook to assess the ecological effects of the military use of chemical defoliants and herbicides." Details of letters and conferences involving the Departments of State and Defense, the National Academy of Sciences, and the AAAS are made available in the study. As the author notes, the assessment of the use of herbicides in Vietnam is "more complex and difficult" than the same task is in the United States predicated by several factors: (1) the controversial nature of U. S. participation in the conflict; (2) the moral issue of using the chemicals; (3) the "informal and limited" nature of the conflict; and (4) the military security imposed on information concerning the hostilities and the use and effects of the herbicides.

The report provides a well-documented historical background of the use of herbicides and defoliants in various guerilla uprisings throughout the world which followed the close of World War II. Public mention of America's use of defoliants in Vietnam was first made in a story in the *New York Times* in late 1961. Planes outfitted with special tanks carrying defoliants flew 60 flights in 1961 and 107 in 1962. Apparently after these first experiments with herbicides, there was a lull in the program while a military assessment of their effectiveness was conducted. In 1962 a State Department senior adviser visited Vietnam and issued a memorandum in which he discussed the use of defoliants and said it "had political disadvantages and was of doubtful benefit." Initially, studies conducted by the Department of Agriculture for Defense's Advanced Research Projects Agency (ARPA) were concerned with the military effectiveness of the chemicals. However, questions were raised in 1965 concerning residues in plants and the soil; but, the report states, "in the initial stages no significant emphasis appears to have been placed on the need for research concerning the long-range effects of herbicides on the ecology." Other Agriculture studies for ARPA included (1) vegetation of Southeast Asia and studies of forest types, December 1965; (2) forests of Southeast Asia, Puerto Rico and Texas (the study emphasized the analogous features of the forests of these 3 areas), September 1967; and (3) response of tropical and subtropical woody plants to chemical treatments, February 1968.

In the spring of 1965, the first program of aerial spraying to destroy food crops was begun in Vietnam. This, too, was covered in a story in the *New York Times*. By early 1968, a program for removal of jungle and crop destruction was well established. According to the report, "the military use of herbicides in Vietnam reached its peak in the fiscal year 1967, and declined somewhat in the following 2 years." Whether or not this can be attributed to an increased outcry from both the scientific and public sectors of society is difficult to assess but there was evidence of a decline in military herbicide usage.

The role of the AAAS in assessing the use of herbicides is examined in detail beginning in 1966 with the resolution submitted by E. W. Pfeiffer, associate professor of zoology at the University of Montana, to the Council of the Pacific Division. Dr. Pfeiffer's original resolution, with emphasis on the use of "chemical and biological warfare agents in Vietnam," received a mixed reception and it was forwarded without recommendations to AAAS's national office. After much debate during the AAAS's annual meeting in Washington, D.C. in 1966, a resolution was passed which, according to the report, dealt primarily with "environmental impairment on a global basis and only very secondarily with military use of herbicides." Also, Dr. Pfeiffer had hoped for an actual scientific field investigation by qualified ecologists under AAAS sponsorship in Vietnam, but this was not included in the resolution. The resolution did "express its concern over the long-range consequences of the use of biological and chemical agents which modify the environment, establish a committee to study such use, including the effects of chemical and biological warfare agents," and volunteer its cooperation to the government "to ascertain scientifically and objectively the full implications of major programs and activities which modify the environment and affect the ecological balance on a large scale."

In March 1967, the AAAS board of directors formed an ad hoc committee on environmental alteration which recommended the establishment of a permanent Commission on the Consequences of Environmental Alteration. The AAAS board of directors decided to assign the general question of the consequences of environmental alteration to the recommended permanent committee and to take on itself responsibility for the next action on the specific question of military herbicide use in Vietnam. The board recognized, however, that "no effective study of the effects of such agents could be carried out in an active theater of war without military or other official permission and sponsorship." As a result of a series of conferences with the director of the Office of Science and Technology, the president of the National Academy of Sciences, and various officials from DOD, a letter was sent by the AAAS president to the Secretary of Defense urging that a study by an independent scientific institution or committee of both the short-and long-range effects of the military use of chemical agents which modify the environment be undertaken." DOD subsequently commissioned the Midwest Research Institute of Kansas City, Mo. to conduct such a study. The MRI report which was rushed through in under 4 months received a mixed reception from both the National Academy of Sciences, which reviewed it, the AAAS, and the public press.

Thus, by the end of 1968 when the AAAS meeting was held in Dallas, the controversy over military use of herbicides in Vietnam was still in a state of flux. Despite considerable technical data released by DOD asserting that no "seriously adverse consequences" resulted from military herbicide usage and assurances by DOD that its assessment program was continuing, the AAAS board of directors did not revise its recommendations urging specific reductions in the program. The AAAS had been unable to enlist the help of the UN in an on-the-spot study; the MRI had been limited to the open literature and involved no onsite investigations. Also, State Department data released in Saigon failed to quiet those who were still insisting on onsite information. Thus, the AAAS board of directors announced that the association "would participate in a study of the use of herbicides in Vietnam." This precipitated another heated discussion within the AAAS council and a resolution was passed expressing agreement with the study but deleting the name Vietnam. By the end of 1969, "it appeared that the AAAS had virtually exhausted its initiatives." However, the association had obtained from both the Departments of State and Defense support for postwar ecological studies of the long-term use of herbicides in Vietnam and assurances from DOD that herbicide assessments would continue. The UN had agreed to sponsor a full-scale international meeting of world scientists on environmental quality.

In summing up, the author observes that "a large federation of scientific societies like the AAAS can provide a valuable forum in which to discuss issues of great public moment" and that military reassessment might well have taken place, but "the persistent expressions of concern from the governing bodies of the AAAS may have helped to make these reassessments more frequent and more searching." However, the author comments that "any issue on which there are both difficult scientific questions and intense political feeling is unlikely to be resolved in the great forum of discussion that the annual meetings of the AAAS produce." And he asks the questions: "Can scientists, any more than other people, compartmentalize their judgment regarding issues they feel strongly about? Can they ignore the political content and address themselves in pure form to the technical?"

Water Pollution

Robert D. Hennigan

Man's control over water is immense, but not so great that he has yet solved one of the major environmental problems of our times—water pollution. Water quality management is a complex, social, and technical system and must be approached as such to provide effective remedial action. The needs and demands of the people will undoubtedly dictate many program goals. In addition, it will be necessary to break down various barriers for such remedial action. (BioScience 19, no. 11, p. 976-978)

Introduction

Water pollution is one of the major environmental problems of our times. As with most other forms of environmental pollution, it has come about because of the industrial-urban growth and development over the past 60 to 70 years, particularly the past 20 years.

Water pollution results when any input into the water cycle alters water quality to the extent that a legitimate use is impaired or lost. Water pollution abatement—or rather water quality management—has social, economic, political, and technical aspects. Consequently, it must be approached as a complex social and technical system requiring an interdisciplinary input if problems are to be accurately defined and effective remedial action taken.

Understanding of the water pollution problem requires a knowledge of:

1) Water quantity and quality relationships
2) The source, type, and volume of pollutants
3) The effect of pollutants on water use
4) The objectives of water quality management
5) The historical background of regulatory efforts
6) Social and economic development and its impact on water quality

From such basic knowledge and background, the contribution of the educational community can be determined in terms of manpower to be trained, problems to be studied and researched, and community education to be undertaken.

Water Resources

The water resource includes water in all its forms, uses, movements, effects, and locations. Rain, hail, snow, ice, fog, atmospheric vapors, soil moisture, surface water, and ground water are all part of the water resource as are streams, rivers, estuaries, ponds, lakes, oceans, wells, and springs.

Water is a renewable natural resource. It is delivered from the atmosphere in the form of rain, snow, hail, fog, and condensation and returns to the atmosphere by evaporation and transpiration. While on the earth, it runs over the ground to lakes, rivers, streams, and oceans and seeps into the ground to be taken up by growing plants to become a part of the ground-water reservoir, eventually discharging also to streams, lakes, or the ocean. There is a continual cycling of water through this system, propelled largely by solar radiation. Precipitation, transpiration, evaporation, ground-water infiltration and discharge, and surface runoff and streamflow vary from place to place and time to time. Furthermore, at any specific location and time, variation is the characteristic pattern to be expected.

Water is one of the essential elements of life. Because of its need and use, water availability is a shaper of civilization and cultures. The rise of cities and burgeoning modern urbanization, such as the Niagara Frontier Region, the city of New York metropolitan complex, and the urban-industrial concentration along the shores of the Great Lakes, is directly related to water availability and development.

In days past, activities and cultures developed where the water could be conveniently put to use. Water location is no longer the constraint it once was; techniques for water development, use, and transpiration have expanded man's control over water immensely.

Man's concern over water includes water quantity; its availability and volume, in time and place; and water quality, its physical, chemical, radiological, and biological condition, which directly affect its availability for use.

Water Quality

Water quality is determined by natural conditions, by man's activities, by land use development and treatment, and most drastically by waste water disposal.

Water found in nature is acid and alkaline, hard and soft, colored and uncolored, highly mineralized and low in minerals, turbid and clear, mildly and significantly radioactive, saline and fresh, saturated and devoid of oxygen, cold and hot, with and without a variety of elements in trace

---

The author is P. E. Director of SUNY Water Resources Center, College of Forestry, Syracuse University, Syracuse, New York. This is the fourth paper from the Pollution Conference, held at State University, Oneonta, N.Y.

concentration, highly productive and unproductive of biological life.

Influencing factors are surface and subsurface geology; geographic and hydrologic conditions; storage availability; the size and nature of waters under consideration; urban and industrial development; other land use; climate, both seasonal and long term; and the physical, chemical, and biological interaction of all these things.

Natural conditions have been and are drastically affected by man's activities. These include development projects which alter stream and lake regimens as noted further on, and waste water inputs, both from point and diffuse sources.

Point sources include municipal sewage treatment plants; sanitary and storm sewer systems; waste outlets from steel mills, petroleum refineries, chemical plants, dairies, food processing plants, paper mills, metal processing and plating works, tanneries, thermal electric power plants, and coke plants; commercial and pleasure boats; irrigation return water; cattle feeding pens and areas; and radioactive processing and use facilities.

The diffused or nonpoint sources are overland runoff from urban, rural, agricultural, forest, and swamp land; fallout from the atmosphere; ground-water discharge to lakes and streams; and sometimes solid waste from dredging, individuals, municipalities, and industries.

The waste water inputs from runoff generally contribute silt, salts, oil, and other deleterious matter from city streets; radioactive material from fallout; sediment from highway construction and urban and farm land erosion; and pesticides, nutrients, and herbicides from land and area application. Point sources sometimes include these elements but, in addition, contribute sewage, phenols, oil, acids, alkalies, heat, solid waste, radioactive material, heavy metals, bacteria and viruses, detergents, salts, floating and settled solids, organic and inorganic material, dissolved solids, biochemical and chemical oxygen-demanding wastes, and toxic and inert material.

The effect of waste loading is dramatic. Water supply quality deteriorates; beaches are closed; the ecological balance is upset; fish and wildlife are killed; areas of lakes and rivers become cesspools; aesthetic sensibilities are offended by floating solids, colors, oil, and odors; recreational use is severely curtailed; nuisances are created; the commercial fishery declines; algae and weeds proliferate; and the mineral content of waters rises.

Interaction

Water quantity and quality considerations cannot be considered unilaterally. They are intimately related. The water supply becomes the used water to be returned as waste water. The impoundment changes the environment and the ecology of the stream or lake. Changing land use from rural to urban changes the stream regimen, frequently resulting in local flood problems and stream pollution because of loss of base flow and increased waste loadings caused by urban runoff. Inter-basin transfer of water alters downstream conditions because of the loss of water. Impoundments may release cold, deoxygenated water high in manganese to the debasement of downstream uses. Peak power releases from hydro projects create surges and rapid changes in stage with consequent shock to downstream areas. Waste water inputs greatly exceed available capacity of unregulated streams in many places, even with the use of advanced treatment methods, particularly in urban areas.

Ideally, quantity and quality considerations are part of a whole or single system and should be considered accordingly.

System Elements

Water resource and quality management involves the physical resource, the people, and the governmental or institutional arrangement.

It is a multi-layered system: the first layer is the geography and hydrology of the locale; the second is the population distribution; the third is the international, state, and local boundary lines setting forth jurisdictional-geographic areas.

The needs and demands of the people will dictate program goals; the physical resource will be the determinant of technological input and facilities or programs needed; and the institutional pattern will identify the agency or instrumentality charged with the responsibility for implementation.

In addition, there are a myriad of other elements such as the legal foundation, economic concerns, and the relationship between adjacent or related areas within levels of government and between the various disciplines concerned with some band of the water resource spectrum. Consequently, the water resource system is not only multi-layered, but it is also marbled, including diverse and, at times, competing and conflicting interests. This is illustrated by the varied objectives of water pollution control.

Major water quality management objectives include:

1) Domestic water supply for public water systems and for facilities and individuals not served by such systems

2) Industrial water supply for manufacturing, food processing, and cooling purposes

3) Agricultural water supply for livestock and for irrigation

4) Water-borne transpiration and navigation

5) Optimal water conditions for fish and wildlife propagation and survival

6) Recreational use of water for boating, swimming, fishing, water skiing, and aesthetic enjoyment

7) Hydroelectric power development and cooling water for steam electric power generation, and

8) Responsible waste water disposal from point and diffuse sources into the environment.

All of these objectives are not totally compatible. One use may conflict with another: limiting recreational use of a watershed to protect public water supply quality; maintaining impoundment or lake levels for recreational purposes, thereby losing storage for stream flow regulation; use of water for irrigation or transfer of water out of a watershed, thereby decreasing downstream flow; increasing the rate of eutrophication by warming up water because of use for cooling purposes, or by addition of nutrients; and loss of fishery and recreational use because of waste water disposal.

The complex interdisciplinary and systematic elements of water pollution are demonstrated by this conference. All of the individual topics, including radiation, pesticides, population, solid waste, heat, and air, are part and parcel of the water pollution package.

Based on the foregoing general understanding, let us address ourselves to the questions of "Where are we?," "How did we get here?," and "What must be done to reclaim and protect the water resource?"

Where Are We?

It is no longer necessary to convince people that serious water pollution problems exist; this is a generally acknowledged fact. Both the popular and scientific press are loaded with articles and stories on the dismal position we find ourselves in. A prime current example is the oil well debacle at Santa Barbara as well as the continuing issues of eutrophication, pesticides, nutrients, and heat.

The waters of the nation and the state have been degraded. There is hardly a lake or a stream that is untouched. Some of the most obvious examples include the Great Lakes, particularly Lake Erie, where accelerated eutrophication brought on by massive waste water inputs has resulted in impairment and loss of many water uses such as fishery and recreational and aesthetic values; the Hudson River; the marine and harbor waters of New York; the ground waters of Long Island; and the local area waters around all metropolitan centers such as the Buffalo River which is a veritable cesspool of sewage and industrial waste. The number of situations that could be listed is truly endless. Naturally, most are familiar with the stories on the pollution of the Mississippi River, the Potomac River, and Puget Sound to name a few. Even the pollution by paper mills of Lake Baikal in the Soviet Union has been documented in the press.

Analysis of the present water pollution situation results in two major conclusions:

1) The water pollution problem is an urban-industrial phenomena, and

2) The great bulk of pollution input is from sewage and industrial waste water.

How Did We Get Here?

The critical water pollution situation developed because of the great urban-industrial growth coupled with a weak regulatory effort. The foundation of water pollution regulation was the prevention of epidemic, water-borne diseases. Water-borne disease was eliminated for all practical purposes by the mid-thirties by water treatment, disinfection, and abandonment of polluted sources. The emphasis was on municipal water supplies, with little attention paid to waste disposal. A resurgence of interest and concern developed after World War II and has intensified since that time.

The unexpected postwar boom brought with it industrial expansion, new industrial processes, detergents, development, and manyfold increases in use of pesticides, fertilizers, and herbicides, increasing urbanization, an exploding population, prosperity, and increased leisure time. Domestic and industrial demands for water started to spiral out of sight as did the demands for water-based recreation, coupled with a parallel increase in the production of waste water. Rising community standards and a greater concern for the quality of life and aesthetic concerns as well as increasing water usage resulted in a public demand that government undertake programs to properly protect, develop, enhance, and reclaim the water resources, particularly as water shortages occurred and pollutional situations worsened.

Urban Growth

The urban nature of much of the water pollution problem has become increasingly apparent.

The great growth of metropolitan areas is well recognized and well documented. The change in the past 50 years has been truly fantastic. In 1900, one out of 20 Americans lived in an urban areas; and in 1968, 14 out of 20 Americans live in an urban area. During this same period of time, the population has doubled, meaning that while the rural population has dropped about 30%, the urban population has increased 2800%. It is estimated that the existing population will double within the next 30 years and some 80% of this will be in urban areas.

Additionally, the per capita use of water has increased four-fivefold over the past 5 decades from 30 to 150 gallons per capita per day. Similar increases have taken place for other water uses—industrial, recreational, etc.

It is in metropolitan areas where over 80% of the people live, provide the support for industrial development, generate the increasing demand for water, and produce most of the pollutant input into the water system.

Another offshoot of this growth is the increased recreational use of the waters. Boat registrations, for example, have increased from about 100,000 in 1960 to 400,000 in 1966.

Cost analyses for water pollution abatement show that over 90% of the needed expenditures of $26 to $29 billion for the next few years are for industrial and municipal waste collection and treatment facilities, again located in urban areas.

It is estimated that the power needed to serve this increasing population and industrialization will increase from 18,000 megawatts to 48,000 megawatts in the next 24 years. This increase in power demand represents 270% of present capability. Practically all of it will be supplied by nuclear steam electric plants since the state's hydropower capability is almost fully developed with the possible exception of some river basin development or off-stream storage for peak power usage.

The increasing usage of water for domestic, industrial, and cooling purposes will produce like amounts of waste water to be treated and disposed of, and the accelerating land urbanization will likewise intensify waste quality and quantity problems, both locally around metropolitan centers and generally throughout the whole water system.

Out of all these changes and demands has come a new frame of reference. This new frame of reference is one which recognizes the limited water resource base, the ever-expanding need and demands, rising standards, and the public insistence that effective action be taken to bring a halt to exploitation and degradation of the water resource. Simply expressed, it constitutes a change of emphasis from, "How much waste can be put into the waters?" to, "How much waste can be kept out of the waters?" and recognition that any approach made be socially, economically, and technically viable.

Action

Barriers to effective action are a sometimes ill-informed public, lack of manpower, limited capability (particularly in metropolitan areas with multiple governments), fiscal resources, ineffective state and federal programs, limited technical knowledge or inability to bring current knowledge to bear on problems, and conflict between various interests.

An effective water quality management program is possible through: (1) system and process changes, both in municipal and industrial water management use, to reduce the volume of water used and subsequent waste water and to reduce the type or amount of polluting substances or materials. Cases in point would be metering of all water services to reduce water consumption, manufacture of paper by a dry process or with a closed system, and the recent change in detergent manufacture to produce a biodegradable product instead of the former stable, persistent material; (2) design, construction, and operation of necessary drainage and sewage collection and treatment facilities and industrial waste treatment facilities; (3) effective control measures to eliminate or reduce input of heat, oil, sediment, pesticides, nutrients, dissolved salts, boat discharges, and solid waste; and (4) impoundments for stream flow regulation to eliminate nuisance conditions to equalize waste dilution and to maintain acceptable environmental conditions.

The problems and needs in water pollution control are world-wide as well as national, state, and local concerns. The nature and seriousness of the problems depend on the particular situation faced and its technical, legal, economic, and social ramifications. However, the common element is the need for: (1) research programs in physical, social, and biological sciences; (2) education and training at the graduate, undergraduate, and technician level to provide needed manpower; (3) a comprehensive, continuously-operating surveillance program to show long-term trends, to pinpoint problems, to evaluate activity, and to help plan future action; and (4) a public education program to develop support for effective action and to counter misinformation about the water resource and water quality management.

Conclusion

At the present time where changing rules, policies, and standards are the order of the day, some bitter battles will be fought between antagonists over water pollution. I think that the following quote from Schlesinger puts this into proper perspective: "Reason without passion is sterile, but passion without reason is hysterical. I have always supposed that reason and passion must be united in any effective form of public action." At the same time, the changing needs and demands must be recognized, accepted, and accommodated.

Population Pollution

Francis S. L. Williamson

If we assume that human population growth is accompanied by strict curbs on environmental abuses, there remains the unresolved problem of population pollution. This is defined as the consequences, mental and physical, of life in a world vastly more populous and technologically more complex than our present one. In such a world the goals of healthy and happy humans, free from malnutrition, poverty, disease, and war, seem convincingly elusive, and the expression of man's full range of genetic potential impossible.
(BioScience 19, no. 11, p. 979-983)

I do not believe that we would still discuss this problem if we did not look hopefully ahead to the technological achievements that may curb--or at least bring to within "tolerable" limits--the tragic, massive, and still-expanding pollution of the air, soil, and water of the earth. This optimism stems from the long overdue consideration of this problem and the implementation of programs dealing with many of its aspects. Environmental pollution, however, is not a priori inexorably linked to human population growth (Daddario, 1968); and we must assume that although numbers of people will increase, the technology now available can and will provide some of the solutions necessary for our health and survival. Additionally, many of us have similar hopes for our less fortunate cohabitants of the earth--those lacking a technology or a freedom of choice. These solutions include the provision in adequate amounts of clean air to breathe, clean water to drink, and clean food to eat. Even if we make the assumption that, as human population growth continues, strict curbs can be simultaneously placed on environmental abuses, we are still confronted with the unresolved problem of population pollution. This I define as the consequences, mental and physical, of life in a world vastly more populous and technologically more complex than the one in which we currently find ourselves. In such a world the goals of healthy and happy human beings, free from malnutrition, poverty, disease, and war, seem to me convincingly elusive. The expression of man's full range of genetic potential is perhaps impossible.

Human Population Growth

I believe there is general agreement among *knowledgeable* men that current trends in the growth of human populations are not only unacceptable but will result in disaster. The current rate of growth, 2% per year (McElroy, 1969), will result in 150 billion people in 200 years. In terms of the time necessary to double the world's population, it represents only about 35 years. As Ehrlich (1968) points out, this "doubling time" has been reduced successively from one million years to 1000, to 200, to 80, and finally to the present 35 years. If the latter rate continues for 900 years, the earth's population will be 60 million people, or about 100 persons for each square yard of the *total* earth surface. Unfortunately, we have no evidence that indicates any lessening of this doubling rate.

Of immediate concern is the world food crisis, a subject dealt with at the Plenary Session of the 19th Annual AIBS Meeting. At that session it was indicated that no efforts are presently being made that would avert global famines by 1985 (McElroy, 1969). Ehrlich (1968) has stated that such famines will be prevalent in the 1970's. What we are prepared to consider as "widespread global famine" is questionable, but apparently it is not the 3.5 million people or more who will starve this year (Ehrlich, 1968); nor is it the general agreement that one-half of the world's people are presently either malnourished or undernourished. I agree with those who feel that we must increase food production at home and abroad in an intensive effort to avert famine, but obviously the successes of such an effort will be pitifully short-lived unless population control is achieved.

While there appears to be general agreement that the growth of human populations must be controlled, both in the long-term sense of allowing for survival and in the immediate sense of averting or alleviating famines, there is little agreement as to how this control is to be achieved.

On 7 January of this year, the Presidential Committee on Population and Family Planning proposed a $120 million increase in the federal appropriation for family planning services to make such services available to all American women who want them. At that time the President stated that no critical issue now facing the world, with the exception of peace, is more important than that of the soaring population. He further stated that world peace will probably never be possible if this latter problem goes unsolved, and he noted that the federal investment in family planning activities had risen from $6 million in fiscal 1964 to

---
The author is Director of the Chesapeake Bay Center for Field Biology (Smithsonian Institution, Office of Ecology), Edgewater, Md.
This is the fifth paper from the Pollution Conference, held at State University College, Oneonta, N.Y.

$115 million in fiscal 1969. This funding may indicate progress, but certainly not of a magnitude proportional to the enormity and urgency of this situation. A value judgment has been made as to the priority of this problem with that of landing a man on the moon, at least regarding funding and the attraction of intellectual effort.

In my opinion the focus continues to be on family planning, not on population control, and this does little more than achieve a reduction in birth rate of an inadequate nature. I find that I must agree with Kingsley Davis (1967) that "There is no reason to expect that the millions of decisions about family size made by couples in their own interest will automatically control population for the benefit of society." As Davis points out, the family planning campaigns in such "model" countries as Japan and Taiwan have hastened the downward trends in birth rates but have not provided population control. Results of the present approach can only be measured as the difference between the number of children women have been having and the number they want to have. For example, the family planning program in Taiwan, assuming that the contraceptives used are completely effective, would be successful if it resulted in the women having the desired 4.5 children each. This represents a sharp drop from the average 6.5 children previously borne to each woman but results in a rate of natural increase for the country of close to 3%. If the social and economic change of Taiwan continues, a further drop in fertility may occur. It may even reach that of the United States, where an average of 3.4 children is currently desired. This would result in Taiwan in a 1.7% per year increase, or a doubling of the population in 41 years, and hardly suggests that our country be used as a model or yardstick for other nations.

The plan of Taylor and Berelson (1968) to provide family planning instructions with maternity care may be a logical step in population control, but I fail to see in what way this plan can alter the basic desire of women in the underdeveloped nations to have more than two children. The natural processes of modernization and education have failed to do this in those nations that are developed. With these facts in mind, it is difficult to imagine the acceptance anywhere that *any* population increase, no matter how trivial, can be tolerated and that the goal must be zero growth.

There is no easy single solution to achieving a zero or near-zero growth rate. Berelson (1969) has recently reviewed the further proposals which have been made to "solve" the population problem. He has appraised them according to scientific, political, administrative, economic, and ethical criteria as well as to their presumed effectiveness. The proposals range from the very nebulous one of augmented research effort to the stringent one of involuntary fertility control. The barriers to acceptance of these criteria, for the truly effective measures, seem insurmountable at the present time. It is my personal view that in the United States a system of tax and welfare benefits and penalties; a liberal, voluntary program of abortion and sterilization (government sponsored and financed, if necessary); attempts at the development in women of substitutes for family interests; and greatly intensified educational campaigns are in order. If the United States is to lead the way, and certainly no other nation appears economically prepared to do so, it seems reasonable that we might begin by abolishing those policies that promote population growth. However, I do not believe we will do these things until economic hardship makes them mandatory. Nonetheless, we are nearing the point of either exercising the free choice of methods of population control still available or facing the compulsory ones that otherwise will be necessary for survival.

In the long interim, however, the emphasis of our efforts can be logically focused on the improvement of the quality of the environment and of the people who are to live here. Here, at least temporarily, we can "do things" with some expectation of success.

The Shift to Urban Life

The rapidly rising number of human beings is not resulting in their general distribution over the landscape but rather in the development of enormous urban centers. In 1800, over 90% of the population of the United States, albeit only some 5-1/2 million people, lived in a rural environment. By 1900, the population of this country was nearly equally divided between cities and rural areas. In 1950, the urban population was 64%; in 1960, 70%; and it is presently about 75%. The projection for 1980 is 78% and for the year 2000 about 85% of the expected 300 million people will be urban dwellers. The number of residents in rural areas has not changed over the last 30 years, and is not expected to vary from its present approximately 53 million persons for the next 10 years. There resides in these data, however, the basic fallacy that what we term "rural" is changing also.

Gigantic urban concentrations are developing within the United States, and these have been termed megalopolises. The three best known have been recently termed "Boswash," "Chipitts," and, for lack of an equally ominous name, "Sag." "Boswash" reaches from New England to Washington, D.C., "Chipitts," from Chicago to Cleveland and south to Pittsburgh and, "Sag," a seaside city occupying the coast of California from San Francisco to San Diego. Demographers and urban planners predict the development of hosts of such "super cities." The Task Force on Environmental Health and Related Problems reported to Secretary Gardner in 1967 that virtually no effort is being made to explore ways of preventing this startling growth. A research program must be inaugurated, they reported, aimed at determining and perfecting measures to shift the focus of future population growth away from already crowded urban areas to parts of the country that are *not now* (emphasis mine) burdened by too many people. Unless such an effort is successful, the pollution control efforts of today, and those planned for the future, could be reduced literally to zero by the sheer increase of people and their correspondingly increased demand for goods, services, and facilities. Similarly, Mayr (1963) earlier pointed out that long before man has reached the stage of "standing room only" his principal preoccupation will be with enormous social, economic, and engineering problems. The undesirable by-products of the crowded urban areas are so deleterious that there will be little opportunity left for the cultivation of man's most uniquely human attributes. This could be what is in store for "Chipitts," "Boswash," "Sag," and others.

It seems that there is not only an urgent need for population control but for planned communities and the de-emphasis of the enormous urban concentrations that compound our problems of coping with environmental pollution.

Urbanization, Pollution, and Man's Welfare

Thus far the data substantiating my remarks are more than adequate. The growth of the world's population is staggering and is attended by increasing urbanization. I still have neglected consideration of man's welfare under these circumstances. The steadily mounting volume of published and unpublished data regarding environmental pollution has focused primarily on the impact of man's activities on his environment and less on the reverse, i.e., the impact of the resultant changes on man himself. Some of these changes are quantifiable, especially those affecting physical well-being. Unfortunately, others affecting such things as mental health and what we refer to vaguely as the "quality of life" are not quantifiable although certainly they are no less real. Obviously, some of these matters cross a number of areas of interest. Consideration of them will be incorporated in the papers of other participants in this symposium.

I would like to consider first some of the quantifiable effects, prefacing my remarks by reiterating the well-known fact that many environmental hazards are so subtle as to be beyond an individual's perception and control. It is less well known that there are frequently some deleterious effects stemming from the most cleverly contrived technological efforts to improve man's general well-being. If we look briefly at selected data from the United States, there is evidence linking air pollution with major respiratory diseases (Task Force on Environmental Health and Related Problems, 1967). Deaths from bronchiogenic carcinoma range from 15 per 100,000 population in rural areas to 30 or more in urban centers with over one million population. Deaths due to emphysema have risen from 1.5 per 100,000 population in 1950 to about 15 in 1964. The correlation of bigger cities with more air pollution with more related deaths seems well-substantiated. Almost half of the people in the United States, 95 million, drink water that is below present federal standards or of unknown quality. Such diseases as infectious hepatitis appear to be directly related to contaminated drinking water, but very little is known about how the agent of hepatitis gets into the water or how it can be removed (The Task Force on Environmental Health and Related Problems, 1967). The concentration of lead is increasing in the air, water, and food, and the blood levels are sufficiently high in many cases to be associated with subacute toxic effects (Dubos, 1965). The accumulation and effects of nonbiodegradable biocides present another serious problem. Documentation is growing that a number of other diseases are associated with environmental pollution, frequently those associated with urbanization.

As alluded to earlier, our best efforts to reduce environmental hazards often have proceeded without adequate knowledge. The development of efficient braking systems for motor vehicles has led to increased exposure of the public to asbestos particles produced by the gradual wearing of brake linings. There is a scientific basis for concern that these particles may promote bronchiogenic carcinoma (Task Force on Environmental Health and Related Problems, 1967). Subsequent to the inoculation of millions of people with a vaccine to prevent poliomyelitis was the discovery that some of the stocks of vaccine, perhaps as many as 25-35%, contained Simian Virus 40 (Sweet and Hilleman, 1960), previously unknown to be resident in the rhesus monkey cells used to culture the virus of poliomyelitis and thus to manufacture the vaccine. The high prevalence of the virus in the cell cultures was compounded by pooling cells from several monkeys. Simian Virus 40 was subsequently shown to be tumorogenic when injected into young hamsters (Eddy, 1962), to possess the capacity of transforming human renal cell lines in vitro (Shein and Enders, 1962), and to result in the production of neutralizing antibodies in 5.3% of cancer patients living in the known limits of distribution of the rhesus monkey (Shah, 1969). Its carcinogenicity in man remains unknown. Poliomyelitis itself is a disease whose spread is enhanced by close human association. Numerous facts support the view that the disease is an enteric infection spread primarily by contaminated excreta (Bodian and Horstmann, 1965).

Without belaboring the matter of pollution and physical well-being excessively, I would like to add that the Food and Drug Administration has estimated that the American people are being exposed to some 500,000 different alien substances, many of them over very long periods of time. Fewer than 10% of these have been analyzed in a manner that might provide the basis for determining their effects, and it has been emphasized that we simply cannot assess potential hazards. The Simian Virus 40 example seems to substantiate this opinion. Nonetheless, severe physical manifestations can ultimately result from repeated exposure to small concentrations of environmental pollutants. These pollutants can have cumulative delayed effects such as cancers, emphysema, and reduced life span (Task Force on Environmental Health and Related Problems, 1967). A three-session symposium of the recent meeting of the American Association for the Advancement of Science was devoted to discussions of such unanticipated environmental hazards, including interactions between contaminants and drugs, food and drugs, and among different drugs.

Earlier I stated my view that environmental pollution and population growth are not inexorably linked. Assuming that our technology renders the environmental scene once again "pristine" in the sense of allowing for sufficient ecosystem function, perhaps even to the point of eliminating the potential health hazards just mentioned, what are the consequences to man's mental well-being of continued population growth and social contacts? If, for instance, we eliminate the dangerous substances in automobile exhausts and asbestos brake linings, how will we be affected by the increase in vehicles from the present 90 million to the 244 million expected to be present in 30 years? We have no data which allow us to establish levels of tolerance for congestion, noise, odor (perhaps removable), general stress, and accident threats, including those from traffic. Excessive exposure to high noise levels can impair hearing or cause total deafness, but the effects of daily noise and disruptions of all kinds, in terms of average human tolerance, is largely unknown (Task Force on Environmental Health and Related Problems, 1967). René Dubos states that: "You can go to any one of the thoughtful architects or urban planners...

none of them knows what it does to the child to have a certain kind of environment, as against other kinds of environments. The whole process of mental development, as affected by physical development of cities, has never been investigated."

I believe it is germane to this discussion to go back to the pioneer work of Faris and Dunham (1939) on mental disorders in urban areas. A brief summary of the data supplied in that study indicates how the incidence of major psychoses are related to the organization of a city. Mental disorders show a decrease from the center to the periphery of the city—a pattern of distribution shown for other kinds of social and economic phenomena such as poverty, unemployment, juvenile delinquency, crime, suicide, family desertion, infant mortality, and communicable disease. Positive correlations are difficult to draw from these data, but they are certainly suggestive and tempting. Each of the chief types of mental disorders has a characteristic distribution with reference to the differentiated areas found within the large, modern city. There is a high degree of association between different types of psychoses as distributed in different urban areas and certain community conditions. It is pointed out that social conditions, while not primary in causation, may be underlying, predisposing, and precipitating factors. Situations involving stress and strain of adjustment may, in the cases of persons constitutionally predisposed, cause mental conflict and breakdown. If social conditions are actually precipitating factors in causing mental illness, then control of conditions making for stress in society will become a chief objective of a preventive program. The study of Faris and Dunham was the first to indicate a relationship between community organization and mental health and to show that urban areas characterized by high rates of social disorganization are also those with high rates of mental disorganization. Finally, it appears that the effect of movement is important to the social and mental adjustment of the person, and precipitating factors in mental breakdown may be found in the difficulties of adjustment to a new situation. Similarly, Dubos (1968) points out that the amount of physical and mental disease during the first phase of the Industrial Revolution had several different causes, one of the most important being the fact that large numbers of people from nonurban areas migrated within a few decades to urban centers. These persons had to make the necessary physiological and emotional adaptations to the new environment.

Our public concern for health, including mental health, has been mainly with frank, overt disease. Since World War II, there has been an increasing understanding of tensions and social stresses, enabling workers in mental health to increase their viewpoints and to include these largely psychologically determined disturbances within their area of interest. The solution of such problems requires the skills of many professions and governmental action nationally and internationally (Soddy and Ahrenfeldt, 1965).

Selection in a Changing World

The concentration of urban life is evidenced by the fact that approximately 70% of our population is crowding into urban areas which represent 10% of the land in the United States. There are presently about 140 million people living on 35 thousand square miles. The evidence reviewed thus far can be reasonably assumed to form the basis for predicting that there will be little or no change in the trend of increasing urbanization. This sequence imposes on man the necessity of ultimately adapting to an environment almost wholly alien to any present today. While our current cities may be no more densely populated than some urban centers have been for centuries, they are infinitely larger and rapidly threaten the existence of all open space. Voluntary population control seems quite unlikely. As long as space and food exist anywhere, it seems reasonable to assume that urbanization will continue until mankind is spread densely over the face of the earth. The luxury of open space appears already threatened and the concept of "getting away from it all" a vanishing one. By 1980, to keep up with today's ratios of people to public space, we will need 49 million acres of national parks, monuments, and recreation areas instead of our present 25 million, and we will require 57 million acres of national forests and 28 million acres of state parks (Task Force on Environmental Health and Related Problems, 1967). It is difficult to speculate what such needs will be in the year 2000, or if at that time it will even be legitimate to consider them as needs.

Dubos (1968) has pointed out that the effects of crowding, safe limits so to speak, cannot be estimated simply from the levels of population density. The populations of Hong Kong and Holland, for example, are among the most crowded on earth, and yet the inhabitants enjoy good physical and mental health. Centuries of crowding have resulted in patterns of human relationships minimizing social conflicts.

The cultural evolution of man from the Neolithic to modern times has taken place without visible biological evolution (Stebbins, 1952). Mayr (1963) points out that Cro-Magnon man differed physically from modern man no more than do the present members of various races one from the other. Crow (1966) views human evolutionary changes as being of such long-term nature as to be considerably less urgent than the problems of increasing population and its relation to natural resources and the quality of life.

Nonetheless, natural selection is important for modern man because it will result in populations of those human beings for whom survival is possible in a uniformly and densely populated world. It is difficult to imagine that time will allow for any considerable shift in man's present genetic makeup, but rather that within the confines of that limitation he must demonstrate the adaptability necessary for continued existence. Such adaptability will necessarily need to be sufficiently flexible to allow for the disappearance of what we now consider basic freedoms and for the increasing regimentation that seems a certain concomitant of future life on earth.

Summary

I would like to summarize by saying that I am optimistic that modern technology can exercise some considerable control over environmental pollution, and that the current ecological crisis in the world makes it seem certain that some progress, perhaps a goodly amount, will be made. I believe that there is less possibility that current trends in the growth of human populations can be changed for a long period of time. Alterations in these trends require changes in the social and cultural fabric of man and society that are linear in nature, while the growth of population numbers is exponential. Family planning is a start, but it must be followed promptly by other programs much more decisive in character. The United States should take the immediate initiative by abolishing all policies promoting population growth and should use its vast economic and intellectual resources to aid in suitable programs elsewhere. Following our earlier and continuing largess in supplying food and medical services abroad, such accompanying programs of aid in population control would seem to constitute a moral responsibility of considerable magnitude.

In the United States efforts must be made to de-emphasize the trend toward huge urban concentrations, to strive for better planned communities, and thus to alleviate simultaneously the problems of pollution and create greater environmental diversification. Predictive technology must be radically increased, and the liberation of substances into the environment curtailed to allow for *at least* a preliminary assessment of effects.

We are presently unable to adequately evaluate those factors influencing mental hygiene in populations and thus to know what the effects of crowding will be on future generations. However, I think it highly unlikely that those people will either think or react as most of us do today. The prospects for continued life as we presently know it seem to me rather remote. Haldane remarked that the society which enjoys the greatest amount of liberty is the one in which the greatest number of human genotypes can express their peculiar abilities. I am apprehensive as to what these genotypes might be, and in what kind of society they will appear, because the complex environment in which man evolved as the most complex biological species is rapidly disappearing. We must realistically face up to the fact that our biological inheritance, in its currently recognizable form, is not going to persist. I agree that to live is to experience, and that to live well we must maintain ecological diversity, a full range of environmental options so to speak, to insure that a wide range of possibilities exist among men (Ripley, 1968). Nonetheless, a full range of environmental options is different things to different people, and survival in a world restricted in options, of a sort alien to me, brazenly confronts mankind.

References

Berelson, Bernard. 1969. Beyond family planning. *Science,* **163**: 533-543.

Bodian, David, and Dorothy M. Horstmann. 1965. Polioviruses. In: *Viral and Rickettsial Infections of Man,* 4th ed., Frank L. Horsfall, Jr., and Igor Tamm (eds.). J. B. Lippincott Co., Philadelphia.

Crow, James F. 1966. The quality of people: Human evolutionary changes. *BioScience,* **16**: 863-867.

Daddario, Emilio Q. 1968. A silver lining in the cloud of pollution. *Med. Opinion and Rev.,* **4**: 19-25.

Davis, Kingsley. 1967. Population policy: Will current programs succeed? *Science,* **158**: 730-739.

Dubos, René. 1965. *Man Adapting.* Yale University Press, New Haven

———. 1968. The human environment in technological societies. *Rockefeller Univ. Rev.,* July-August.

Eddy, B. E. 1962. Tumors produced in hamsters by SV40. *Fed. Proc.,* **21**: 930-935.

Ehrlich, Paul R. 1968. *The Population Bomb.* Ballantine Books, Inc., New York.

Faris, Robert E. L., and H. Warren Dunham. 1939. *Mental Disorders in Urban Areas.* University of Chicago Press, Chicago.

Mayr, Ernst. 1963. *Animal Species and Evolution.* Harvard University Press, Cambridge.

McElroy, William D. 1969. Biomedical aspects of population control. *BioScience,* **19**: 19-23.

Ripley, S. D. 1968. Statement in Joint House-Senate Colloquium to Discuss a National Policy for the Environment. Hearing before the Committee on Interior and Insular Affairs, United States Senate and the Committee on Science and Astronautics, U.S. House of Representatives, 90th Congress, 2nd Session, 17 July 1968, No. 8, p. 209-215.

Shah, Keerti V. 1969. Investigation of human malignant tumors in India for Simian Virus 40 etiology. *J. Nat. Cancer Inst.,* **42**: 139-145.

Shein, H. M., and J. F. Enders. 1962. Transformation induced by Simian Virus 40 in human renal cell cultures. I. Morphology and growth characteristics. *Proc. Nat. Acad. Sci.,* **48**: 1164-1172.

Soddy, Kenneth, and Robert H. Ahrenfeldt (eds.). 1965. *Mental Health in a Changing World.* Vol. 1 of a report of an international and interprofessional study group convened by the World Federation for Mental Health. J. B. Lippincott Co., Philadelphia.

Stebbins, George L., Jr. 1952. Organic evolution and social evolution. *Idea Exp.,* **11**: 3-7.

Sweet, B. H., and M. R. Hilleman. 1960. The vacuolating virus, SV40. *Proc. Soc. Exp. Biol. Med.,* **105**: 420-427.

Taylor, Howard C., Jr., and Bernard Berelson. 1968. Maternity care and family planning as a world problem. *Amer. J. Obstet. Gynecol.,* **100**: 885.

The Task Force on Environmental Health and Related Problems. A Report to the Secretary of Health Education, and Welfare. 1967. U.S. Govt. Printing Office, Wash., D.C.

Research and Development for Better Solid Waste Management

Andrew W. Breidenbach and Richard W. Eldredge

The problem of U.S. solid waste management can now be quantified in terms of waste generated in lb/person/day. This development of reliable data is part of an effort initiated and then accelerated by the 1965 Solid Waste Disposal Act, the legislation that responded to national needs for new solid waste management policies and technology. The thrust of research and development has been organized into a R&D matrix. Discussion of the six segments of this matrix and some of the projects associated with it provide an introduction to the national research and development program under authority of the Act. (BioScience 19, no. 11, p. 984-988)

Preliminary data from a national survey just completed by the Bureau of Solid Waste Management (Muhich et al., 1968) shows that in 1967 for each man, woman, or child, 10 lb. of household, commercial, and industrial solid wastes were generated. The total was over 360 million tons per year, and this figure did not include 550 million tons per year of agricultural wastes and crop residues. Neither did it include approximately 1.5 billion tons per year of animal wastes, nor over 1.1 billion tons per year of mineral wastes. In grand total, it was estimated that more than 3.5 billion tons of solid wastes are being generated in this nation every year (Black et al., 1968).

Convinced of the need for basic and applied research in order to manage these solid wastes, the 89th Congress enacted the Solid Waste Disposal Act of 1965 (PL 89-272). Passage of the Act initiated and then accelerated a national program of research and development, which included development of adequate and reliable data such as that coming out of the 1968 National Survey of Community Solid Waste Practices. The national research program was aimed at finding new and improved methods of effective and economic solid waste management.

Dr. Breidenbach is Director, Division of Research and Development, and Mr. Eldredge is Director, Program Development, Bureau of Solid Waste Management, Environmental Control Administration, Consumer Protection and Environmental Health Service, Public Health Service, U.S. Department of Health, Education, and Welfare.
This is the sixth paper from the Pollution Conference, held at State University College, Oneonta, N.Y.

The Act directed the Secretary of Health, Education, and Welfare to conduct research in solid waste management, to encourage, to cooperate with, and to render assistance to, public and private agencies and institutions as well as to individuals in the promotion of solid waste research and development. In addition, one of the intents of the Act was to encourage conservation of natural resources by reducing waste and unsalvageable materials, and by recovering the potential resources hidden in the mounting discards of our civilization. The Department of the Interior administers the Solid Waste Disposal Act as it relates to solid wastes resulting from the extraction, processing, or utilization of minerals or fossil fuels within those industries. The Department of Health, Education, and Welfare (HEW) administers the Act related to wastes from all other sources.

Operating within the legislative authority of the Solid Waste Disposal Act, HEW's Bureau of Solid Waste Management has set forth goals to aid the nation in achieving effective solid waste management as follows: (1) to protect man's health and the quality of his environment; (2) to minimize the amounts of ultimate waste being generated; (3) to maximize the salvage of useful waste material.

The Bureau's program is headquartered in Rockville, Maryland, as part of HEW's Environmental Control Administration. The Bureau's Research and Development Division laboratory and pilot-plant facilities are maintained in Cincinnati, Ohio, and at Johnson City, Tennessee. The Bureau's mission is carried out through three divisions: the Division of Research and Development; the Division of Technical Operations; and the Division of Demonstration Operations.

Although the preliminary data from the National Survey are helping to quantify the problem associated with solid waste management in the United States, as citizens, our individual buying habits also help to characterize the situation we face. Quite naturally we have become accustomed to the use of the word *disposal*. It is simple enough to describe how we acquire goods through the act of purchasing. It is, however, rather difficult to describe exactly how we actually dispose of these goods. Even Webster has difficulty in bringing into lucid terminology the meaning of disposal. We find phrases such as "transferring into a new place" or "to get rid of" or "to put out of the way." The solid matter that is cleared out of our households, our industries, our agricultural enterprises, and our commercial institutions may seem to be disposed of when we can no longer see it. Nevertheless, from an objective viewpoint, this solid matter continues to exist in various shapes—in various locations—for decades before its atoms and molecules become the building blocks of new substances that may ultimately again enter the consumer market and then again the waste stream.

We have watched the increased acceptance of single-use items in the soft drink industry, in hospitals, and in other enterprises. We have noted our increasing dependence upon plastics for a variety of attractive, useful packaging. We watch the discard of tons of refined metal—a practice unthinkable two to three decades ago. There is no question, however, that these advances in technology are examples of a greater era of national productivity. Our prosperity and affluence produce solid waste, and this is a brief description of the problem.

We have recognized that there are at least four identifiable classifications of solid waste: municipal; industrial; agricultural; and commercial. Although we will speak principally about municipal solid wastes, these categories give an idea of the extent of the problem.

We believe that R&D efforts for better solid waste management need to be organized so as to attack the problem in segments that would form the basis for a research and development matrix (Fig. 1). A discussion of the segments of this matrix and brief descriptions of some of the projects associated with it provide an introduction to the national research and development program under authority of the Solid Waste Disposal Act.

Source Reduction

All solid waste has a distinct point of origin; that is, the location where the substances are discarded. It follows that this point of origin, if accurately defined, may provide a key for the modification of some of our present solid waste patterns. Source reduction encourages the concept of decreasing the amounts of solids entering the waste stream at the very point of generation. We are all aware of items which become wastes soon after we acquire them. They appear as boxes, cases, wrappers, bags, bottles, cans, envelopes, and a myriad of other items. Thoughtful consideration, education, and decisions are required if we are to stabilize or reduce the amount of solid wastes generated per capita. However, the effort, ambition, and ingenuity that are applied to the design of the attractive packaging of our time is a credit to those who compete for our various consumer markets. The packaging industry should be encouraged to apply the same level of ingenuity and innovation to the design of packaging that might be readily reclaimed or, at least, that might lessen the solid waste management problem. Some groups have begun action in this direction.

Storage

Every producer of solid wastes must provide a method of holding or storing the wasted material while awaiting collection. The housewife with a food grinder is perhaps the only exception, in that the grinder eliminates the problem of storage. The storage system employed has a direct effect upon the environment of the area in which the waste is produced. Poor storage of solid wastes provides sustenance for rodents and insects, produces unpleasant odors, provides fuel for fires in and around the storage areas, and prevents the efficient and economic collection of the solid waste materials. All aspects of solid waste management may have their effect upon the environment, but storage and collection are more nearly related to the solid waste problem as viewed by the average citizen. This problem literally is at the citizen's very door. Inadequacies affect him directly, quickly, and with great impact.

It should be noted that solid waste storage—except for larger containers to serve larger waste generation—has not changed appreciably since man found it necessary to remove his waste from his immediate environment. The static storage system has perpetuated a collection system, for the most part, of similar antiquity.

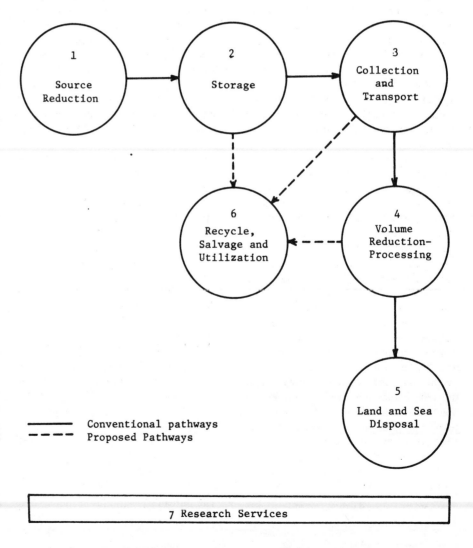

Fig. 1. Research and Development Program Matrix.

Collection and Transport

The collection and transport of solid wastes is a connecting link between storage and processing or disposal and is affected by each of these. Poor storage and illogical disposal practices adversely affect the collection process. It is estimated that 70 to 80% of the cost of solid waste management is accounted for by the collection and transport aspect. This portion of the service has been recognized as *elimination* or *disposal* of waste by the citizen. He has demanded rapid removal of solid waste from his property—considering the job complete when the truck has disappeared around the corner. It seems natural, then, that funds for collection services have been more obtainable than those for ultimate disposal sites. Funds, however, have been used mostly for more trucks, bigger trucks, and increases in manpower. New techniques for collection, in combination with improved storage, are necessary to management of the burgeoning solid waste problem. Land that is suitable for disposal of solid wastes is rapidly being pre-empted for other purposes. The people of suburbia naturally resent the placement of disposal sites nearby. The noise and impact of concentrated vehicular traffic poses additional problems in area and route selection. For these reasons and many more the transportation of solid wastes is becoming increasingly expensive and time-consuming. A comprehensive evaluation of collection and transport, evaluated as a system, can often produce economics of operation and design so that service can be improved without attendant degradation of the environment. Investigations conducted within the recent National Survey of Community Solid Waste Practices show that for 14% of the communities reporting, both separate and combined collection are used sporadically and unpredictably, an inconsistency not in keeping with economic operation. The survey also indicates that over 12% of the citizens are not served by any collection system. Improvements in collection and transport systems technology should provide these areas of economic and environmental conflict with improved service systems.

Volume Reduction Processing

With the increase in urban land utilization, the scarcity of land disposal areas for solid waste has become a serious concern. Sanitary landfill, the technique of the engineered disposal of solid wastes on land, reduces the volume of wastes as a function of normal operation. Any process or system that will further reduce the volume of wastes is extremely desirable. Such a process could add years of additional service to existing landfill sites, reducing the need for future sites. Volume reduction processing systems, such as incineration and composting, whether or not by-products of the operation are fully exploited, may be the method preferred for decreasing refuse volume. Such systems at the present time may produce incidental pollution, thus diminishing their value to the community. Air pollution, water pollution, and vector problems all too often accompany attempts at volume reduction. Although present systems may be employed without degrading the environment, often the cost of protection precludes their use. Studies of existing methods and innovations in similar techniques are sought to provide economic and adaptable concepts in volume reduction methods to preserve land areas. The 1968 National Survey of Community Solid Waste Practices indicated that less than 9% of the citizens are served by incinerators, the major method of volume reduction prior to disposal. Presently, more than 80% of these incinerators are insulting the environment with air pollution, water pollution, visual blight, and vector proliferation.

Land and Sea Disposal

Approximately 6% of the nation's land disposal sites are acceptable by present environmental standards. The open, sometimes burning, dump is the rule rather than the exception. New and better methods of solid waste disposal upon the land must be developed to preserve our environmental integrity. Presently acceptable methods must be evaluated and adopted by even the smaller governmental entities. Cost-benefit methods must be applied to the utilization of land as well as water and air. But as more stringent regulations for land disposal techniques are applied, there is a tendency for government and industry alike to seek other areas of ultimate disposal. Communities and industries in coastal areas view the sea as a natural sink for solid wastes. Because of the immediate and continuing pressure to use the sea, we need factual data on the extent of sea disposal priorities. We must consider, from a well-informed point of view, the consequences of sea disposal upon the marine environment and the ecology of the sea.

Reclamation

Several factors justify projects in the reclamation category: (1) The volumes and weights of solid waste requiring ultimate disposal can be decreased by removing portions of the wastes for salvage, reuse, or recycling; (2) some value does remain in the materials heretofore discarded; and (3) the basic raw materials used in the production of goods are lost to us permanently when solid wastes are disposed of by the methods presently applied. Pertinent salvaging techniques are generally too costly to provide the incentive of profit or a marginal cost return. Conservation of our national resources, including land areas for disposal and reclamation of despoiled land, may be realized through the development of adequate mechanical sorting and classifying devices. New methods of utilization made possible by innovations in recycling and reclaiming materials present in the waste stream may prove advantageous. Regardless of the location of the resource, recycling, or reclaiming operation, materials removed at any time will ultimately reduce the final disposal quantity. This factor alone may prove timely as our national production rate of wastes is increasing. Naturally, one of the first sources of recyclable material is any process that produces a large quantity of waste products which are of uniform, consistent quality. As separation and classification techniques become more sophisticated and efficient, materials intimately mixed with other wastes may assume new economic importance. The recycling and reuse of our natural resources is a basic tenet, held by all, yet unfortunately followed by few. Present attempts at these reuse methods have been limited—usually to indiscriminate hand-picking by unfortunate individuals who are looking for more satisfactory employment. Once solid waste is received at a disposal site, only a small portion is recycled.

Research Services

Since most solid waste is unusable or unwanted, there has been a normal reluctance to identify quantities and sources or to develop methods of characterizing, in any meaningful way, the discards themselves. Before success can be obtained in improving solid waste management, many basic questions must be answered. A few of these are:

1) What are the normal characteristics of solid waste?
2) How can representative samples of such a heterogeneous mixture be made?
3) What analytic methods should be used to define these characteristics?
4) Do pesticides and other toxic materials persist in solid wastes?
5) What are the potential health hazards in a given solid waste management system?

The answers to these and other questions are of concern to all engaged in research and development within the Bureau of Solid Waste Management. Studies in this segment of research and development are designed to provide the basic tools which are necessary to solve the problems represented in the other segments of the matrix.

Attacking Our Problem

There are four methods of accomplishing research and development that can be employed under the Solid Waste Disposal Act of 1965. These can be listed as: in-house effort, contract effort, research grant contributions, and demonstration grant activities. The usefulness of these modes of funding—within a mission-oriented, matrix-directed program—depends upon the urgency, scale, complexity, and resource requirements of the tasks to be accomplished.

Intramural Research

Topics for intramural exploration have a high degree of flexibility, because control of project direction is continuously close to the work itself. Success with intramural work is dependent upon the ability to acquire competent investigators and managers, as well as suitable space and equipment within a reasonable time period. Intramural research must be supported by top-flight personnel recruitment and procurement services.

As an example of an intramural research project, let me tell you of our project to design and test a high-temperature incinerator for small-sized population units. The project is directly associated with our needs for new concepts in refuse incineration that will present alternatives to conventional incineration techniques. In the current phase of the project, the design of the incinerator is complete, and procurement of necessary parts is in progress. It is anticipated that the incinerator will soon be assembled and testing will begin. The potential benefit of this project, if successful, will be an adequate means of incineration for communities in the 10,000-to 50,000-population bracket.

Another project has as its objective definition of the microbiological quality of the total effluent and immediate improvement of current- and to-be-developed incineration processes. Microbiological data are being taken to form baseline information for future use in assessing the efficacy of new and modified incineration processes. Five incinerators have now been tested, and we anticipate that in another year we will have the information from a sufficient number of sites to establish the needed baseline information.

Contract Research

Contract research minimizes the space and personnel considerations inherent in intramural research and permits mutual agreement as to what is to be accomplished before the contract is negotiated. Such research requires close supervision by a knowledgeable and, in most cases, senior scientist or engineer. The contract is a legal document, and those engaged in contract research are looked upon as ex-officio members of the intramural research team. The variety and extent of our research contracts are promising signs of a coming expansion of solid waste technology and fruitful business-government research relationship (Clemons and Black, 1969).

An example of contract research is the subscale experiment program for the Combustion Power Unit-400. These experiments are a planned follow-up to feasibility studies completed during the last fiscal year (Aerospace Commercial Corporation, 1969). This study showed that it may be economically and technically feasible to use the waste heat from the controlled fluid-bed incineration of municipal solid waste to generate electricity with the aid of a gas turbine. The subscale experiments will permit evaluation of the new principles in combustion as well as particulate reduction necessary to the successful implementation of this concept. By the end of this fiscal year we expect to have the information necessary to permit a decision on the construction and testing of a prototype power-generating incinerator. Success with this particular project would result in substantial reduction of costs in volume reduction by incineration.

A second example of contract research is a study of the relationship between packaging materials and waste disposal. This study determined the present proportions for types and volumes of packaging materials and indicated trends to the year 1976, with anticipated effects on solid waste management problems. Means of making changes in packaging to mitigate such problems were suggested (Darnay and Franklin, 1969). The work accomplished is the necessary first step toward the objective of redirecting these materials away from the waste stream and thus reducing the amount of waste remaining to be managed. With the basic data available from this effort, it should be possible to initiate definitive studies and move further toward the objective of minimizing solid waste management problems associated with packaging materials.

Research Grants

This mode of research provides the investigator with a high degree of freedom. The grantor has but modest control over the direction of the research, once the funds are awarded. Reports and publications are related to the freedom of the investigator and are not counted on in the same manner as in a time-sequenced research and development program. Research grants can provide excellent opportunities for exploratory research of high-risk concepts where some initial free-lance investigative efforts minimally related to a time-sequenced matrix are desirable. Thus, the Bureau of Solid Waste Management supports a wide variety of research projects through the grant mechanism (Lefke, 1968).

An example of research grant effort is that being conducted at the University of Pennsylvania on pipe transport of domestic solid wastes. Objectives are to investigate the application of known technology of solid transport in pipes for the collection and removal of solid waste as well as economic comparison with truck collection systems. This basic research on a new and radically different collection system has potential application not only in future model cities but may well be feasible to replace existing collection systems in established cities, if the cost of this installation is amortized over a 50-year period.

Another example of effort being conducted through the research grant mechanism is the grant entitled "Pyrolysis of Solid Municipal Wastes." This work, being performed by the city of San Diego, has as its objective the investigation of the feasibility of pyrolysis as an economic method of decreasing the volume of solid wastes, the production of useful by-products, and the determination of the optimum conditions for operation of the process.

Characterization studies have been made to form the basis of pilot-plant charge materials. The municipal solid wastes are being pyrolyzed at various temperatures, and the resulting solid, liquid, and gas products collected and analyzed. Typical samples have been pyrolyzed at temperatures of 90 F, 1200 F, 1500 F, and 1700 F, using a sample density of 5.55 lb./cu ft. The products to date consist of gases, liquid pyroligneous acids, and somewhat contaminated charcoal.

Demonstration Grants

Although demonstration grants are not organizationally a part of research and development within the Bureau of Solid Waste Management, the demonstration grant mechanism often provides the testing ground for research ideas that have succeeded at the bench- or pilot-plant level, and fosters the testing of full-scale solid waste management facilities by grantees such as communities or nonprofit organizations. The demonstration and testing of such facilities results in the development of information which could be considered for application by other communities. To date, the Bureau of Solid Waste Management has funded 90 demonstration projects, and summaries of projects funded through June 1968 have been published.

An example of the demonstration grant mechanism is the grant entitled "Investigation of the Potential Benefits of Rail Haul as an Integral Part of Waste Disposal Systems." This grant, made to the American Public Works Association, has as its objective the comprehensive evaluation of the costs and benefits obtainable from the collection, transportation, and disposal of solid wastes from urban areas by the use of rail haul techniques. The project was to be conducted in three phases, each of which was expected to take approximately a year to complete. Phase One was concerned with the identification, development, and setting up of the rail haul and related wastes transport and disposal techniques. Phase Two will deal mainly with implementation of the rail-haul waste disposal concept in cooperation with select communities. Phase Three is planned as a comprehensive evaluation of the concept and techniques as demonstrated under actual operating conditions during Phase Two. An interim report covering work done during the first phase has been prepared and will be available from the Bureau. Groundwork for the system has been thoroughly established, and a substantial variety of system elements has been examined. A number of promising alternatives have been identified on which work is being concentrated. Overall, the work to date suggests that the rail-haul concept has considerable promise for alleviating urban solid waste disposal problems. There is a strong likelihood that large savings in the solid waste collection costs might accrue to participating communities. Another potential benefit of rail hauling is the possibility of reclaiming large land areas which are otherwise useless.

Another example of a demonstration project is a full-scale compost plant, located at Gainesville, Florida. Wastes from the City of Gainesville are processed in this plant in an attempt to demonstrate the reliability, suitability, economic feasibility, and the sanitary and nuisance-free operation of a newly developed high-rate mechanical composting system. This system was specifically designed for disposal of municipal refuse from a medium-sized community.

It might be noted that as a parallel effort in our intramural research, we are investigating windrow composting in conjunction with the Tennessee Valley Authority at Johnson City, Tennessee. The Bureau of Solid Waste Management expects to have its first position paper on the efficacy of composting as a means of municipal waste disposal available during the early part of fiscal year 1970.

It is hoped that through research and development we may devise new and improved technology which will help in the management of a variety of solid wastes generated within the United States. However, these techniques cannot effectively mitigate the present problem of poor solid waste management unless four additional elements can be assured: (1) increased awareness and concern of the average citizen for his individual, community, and corporate solid waste management problems; (2) cooperative regional and community action—through professional leadership—to manage solid wastes effectively; (3) the efforts in college and university of faculty and students, who possess the ingenuity and innovative expertise, to bring about new solutions; and (4) the well-known capability of the industries that form the backbone of American technological progress. Thus, if the citizen, the community, the university, and industry will help to create and to test a new technology, the millions of tons of wastes generated each year can perhaps be channeled, used, recycled, managed, and transformed into millions of tons of American assets.

References

Aerospace Commercial Corporation 1969. Combustion Power Unit—400; a technical abstract, U.S. Department of Health, Education, and Welfare, Cincinnati. In press.

Black, R. J., et al. 1968. The national solid wastes survey; an interim report. U.S. Department of Health, Education, and Welfare, Cincinnati, 53 p.

Clemons, C. A., and R. J. Black. 1969. Summaries of solid wastes program contracts; July 1, 1965—June 30, 1968. Public Health Service Publication No. 1897. U.S. Government Printing Office, Washington, D.C. In press.

Darnay, A. J., Jr., and W. Franklin. 1969. The role of packaging in solid waste management, 1966 to 1976. Public Health Service Publication No. 1885. U.S. Government Printing Office, Washington, D.C. In press.

Lefke, L. W., Jr. 1968. Summaries of solid wastes research and training grants, 1968. Public Health Service Publication No. 1596. U.S. Government Printing Office, Washington, D.C. 48 p.

Muhich, A. J., et al. 1968. Preliminary data analysis; 1968 national survey of community solid waste practices. Public Health Service Publication No. 1867. U.S. Government Printing Office, Washington, D.C. 483 p.

Sponagle, C. E. 1968. Summaries; solid wastes demonstration grant projects—1968. Public Health Service Publication No. 1821. U.S. Government Printing Office, Washington, D.C. 89 p.

Thermal Pollution

LaMont C. Cole

There are sources of energy that man can use without raising the average temperature of the earth, but the use of fossil fuels or nuclear or thermonuclear reactions will raise the temperature. This has implications both direct and indirect for the earth's biota. (BioScience 19, no. 11, p. 989-992)

Introduction

There are entymological objections to the expression "thermal pollution," but it is gaining adoption and I shall here accept it without comment as a descriptive term for unwanted heat energy accumulating in any phase of the environment.

Any isolated body drifting in space as the earth is must either increase continuously in temperature or it must dispose of all energy received from external sources or generated internally by one of two processes. The energy may either be stored in some potential form or it must be reradiated to space.

In the past the earth has stored some of the energy coming to it from outside. This was accomplished by living plants using solar energy to drive endothermic chemical reactions, thereby creating organic compounds some of which were ultimately deposited in sediments. This process had two very important effects; in protecting the organic compounds from oxidation it created a reservoir of oxygen in the atmosphere; and it also eliminated the necessity for reradiating to space the heat energy that would have been released by the oxidation of those compounds. Part of the stored organic matter gave rise to the fossil fuels, coal, oil, and natural gas, which we are so avidly burning today—and now the earth must at last radiate that heat energy back to space, or its temperature will increase.

Many of the most important problems currently facing man are ecological problems arising from the unrestrained growth of the human population and the resultant increasing strains being placed on the earth's life support system. Our seemingly insatiable demand for energy is one important source of strains on the earth's capacity to support life, and I propose here to examine it in very elementary terms.

The Earth's Sources of Energy

Well over 99.999% of the earth's annual energy income is from solar radiation and this must all be returned to space in the form of radiant heat. This has gone on forever, so to speak, and conditions on earth, since before the advent of man, have reflected the necessity of maintaining this balance of incoming and outgoing radiation. A ridiculously small proportion of the sunlight reaching us is used in photosynthesis by green plants. This energy powers all of the earth's microorganisms and animals and is converted to heat by their metabolism. In addition, man can obtain useful heat energy by burning organic matter in such forms as wood, straw, cattle dung, and garbage and other refuse. This is "free" energy in the sense that the earth would have reradiated it as heat even if man had not obtained useful work from it in the meantime. Physical labor performed by man and the work done by domestic animals are also means of utilizing this solar energy.

In addition, it is solar radiation that keeps the atmosphere and hydrosphere in motion. To the extent that man can utilize the energy of the winds, of falling water and of ocean currents, or can make direct use of sunlight, he can do so without imposing increased thermal stress on the total earth environment.

The earth has some other minor sources of energy that contribute to its natural radiation balance. The tides can be used to obtain useful energy and a plant in France is now producing electricity from this source. Also, tidal friction is very gradually slowing the rotation of the earth, thereby converting kinetic energy to heat which plays a small role in maintaining the earth's surface temperature before it is radiated to space. Heat is also emerging from the interior of the earth and current concepts attribute this heat to natural radioactivity. In some local areas this heat flux is concentrated, and in a few countries man utilizes it for such purposes as heating buildings and generating electricity. Italy generates more than 400,000 kw of electricity from geothermal heat and in the United States about 85,000 kw are derived from a geyser field north of San Francisco.

This then is the inventory of energy sources available to man without affect-

---

The author is a professor in the Division of Biological Sciences, Cornell University, Ithaca.
This is the seventh paper from the Pollution Conference, held at State University College, Oneonta, N.Y.

ing the surface temperature of the earth or the quantity of heat energy that it must dispose of by radiation. It is important to note this because, at least in the "developed" nations, we actually seem to be regressing from the use of these natural energy sources. Certainly windmills, animal power, and manual labor are much less in evidence today than they were during my childhood, and useful work done by sailboats seems to be a thing of the past in our culture. When we burn fossil fuels or generate electricity by nuclear or thermonuclear reactions, we must inevitably impose an increased thermal stress on the earth environment and, to the extent that this heat is undesirable, it constitutes thermal pollution.

The Earth's Radiation Balance

I assume the mean temperature of the earth's surface to be 15 C or 288 K; this may be a degree too low or too high. In order to keep the discussion sufficiently simple-minded, I shall introduce three simplifications.

First, I shall treat the earth as though its entire surface was at a uniform temperature equal to its average value, thus ignoring temperature differences due to latitude, altitude, season, and time of day. The effect of this simplification is less drastic than one might expect. It is true if one measures radiation from several bodies at different temperatures and infers the mean temperature of the surfaces from the radiation, he will obtain values that are slightly too high. For example, if we have two otherwise identical areas, one at 0 C and the other at 50 C, the average of their combined radiation will equal that which would be given off by such an area at 28 C rather than at the true mean temperature of 25 C. For the earth as a whole, the error from this source is not likely to amount to even one degree, and the other two simplifications I shall make tend to cause small errors in the opposite direction.

Second, I assume that the earth radiates as a blackbody or perfect radiator. This is probably nearly correct; and, in any case, the assumption is conservative for our purposes here. If the earth is actually a gray body rather than a black one, it will have to reach a somewhat higher temperature to dispose of the same amount of heat energy.

Finally, I assume that the earth radiates its energy to outer space which is at a temperature of 0 K. This is not quite correct because the portions of the sky occupied by the sun, moon, stars, and clouds of interstellar matter are at temperatures above absolute zero, and the earth's surface must therefore be very slightly warmer than it would otherwise be.

Accepting these simplifications, we can now turn to the Stefan-Boltzmann law of elementary physics which states that radiation is proportional to the fourth power of the absolute temperature, and easily calculate the amount of energy radiated from the earth to space.[1] At a temperature of 15 C (288 K) the total radiation from the earth turns out to be 2×10^{24} ergs/sec.

Or, looked at the other way around, if we know the quantity of heat that the earth's surface must get rid of by radiation, we can calculate what its surface temperature must be for it to do so. For example, various students of the subject have concluded that the mean temperature difference between glacial and ice-free periods is quite small, probably no more than 5 C (e.g., see Brooks, 1949). If we can assume that we are now about midway between these climatic types, then a rise of 3 C might be expected to melt the icecaps from Antarctica and Greenland thereby raising sea-level by some 100 m. This would drastically alter the world's coastlines as, for example, by putting all of Florida under water, and drowning most of the world's major cities. To do this would require a 4.2% increase in the earth's heat budget (an increase of 8.44×10^{22} ergs/sec). By the same reasoning, a 4.1% decrease in the energy budget could be expected to bring on a new ice age. As we shall see presently, however, these things are not really quite so simple and predictable.

Man's Effect on the Heat Budget

The amount of energy now being produced by man, I take to be 5×10^{19} ergs/sec or 25/1000 of 1% of the total radiated by the earth. Strangely enough, I am

---
[1] I take the surface area of the earth as 5.1×10^{18} cm² and the value of the Stefan constant as 5.67×10^{-5} erg cm^{-2} deg^{-4} sec^{-1}

quite happy with this perhaps brash attempt to estimate a difficult quantity. Putnam (1953) estimated the fuel burned by man 16 years ago at 3.3×10^{19} ergs/sec. If he was correct then, and if energy demand then had been growing at the rate it is now growing, man would now be using twice what I have estimated. Kardashev (1964) estimates the energy now produced by civilization at "over 4×10^{19} ergs per second" which is consistent with my estimate.

There is another way of getting at the figure. The rate of combustion of fossil fuels in the United States is accurately known, as is our rate of generating electricity. Our electrical production as of 1966 corresponds to one-fifth of our fossil fuel consumption. If we assume the same ratio for the entire earth, for which we do have credible figures on electrical generation, the world energy production turns out to be 4.4×10^{19} ergs/sec. With this many independent estimates converging, I am happy with the figure of 5×10^{19} ergs/sec.

This is such a tiny part of the earth's output of energy that it is evident that the heat released by man now has an absolutely insignificant direct effect on the average temperature of the earth's surface. Will this be true if we go on increasing our demands for power? I have been hearing utility company officials assert that we must keep electrical generating capacity growing by 10% per year, but a more common projection is about 7%, which rate the capacity would double every 10 years. I am confident that nonelectrical uses of fuel for such purposes as heating, industry, and transportation are growing at least as rapidly, and I am told that the "developing" countries are going to continue to develop. Let us examine the consequences of an energy economy that continues to grow at 7% per year.

Waggoner (1966) considers it at least possible that a warming of 1 C would cause real changes in the boundaries between plant communities. For this to occur, the earth's energy budget would have to increase by about 3×10^{22} ergs/sec. How long would it take man to cause this at an increased energy production of 7% per year? The answer is 91 years.

As already mentioned, a rise of 3 C could, in the opinion of competent authorities, melt the ice caps and produce an earth geography such as has never been seen by man, and on which there would be much less dry land for man. This would take about 108 years to achieve by present trends. Let us rejoice that the people during our Civil War did not embark on an energy-releasing spree as we are now doing.

The highest mean annual temperature of any spot on earth is believed to be 29.9 C (302.9 K) at Massawa on the Red Sea in Ethiopia. I think it is safe to assume that if the average temperature of the whole earth was raised to 30 C, it would become uninhabitable. This would take 130 years under our postulated conditions—the time from the Victorian Age to the present.

These calculations, rough as they are, make it clear that man is on a collision course with disaster if he tries to keep energy production growing by means that will impose an increased thermal stress on the earth. I have here ignored the fact that the fossil fuels, and probably uranium and thorium reserves also, would be exhausted before these drastic effects could be attained. The possibility exists, however, that a controlled fusion reaction will be achieved and bring these disastrous effects within the realm of possibility.

How can we avoid the consequences of the trends we are following? My answer would be to determine what level of human population the earth can support at a desirable standard of living without undergoing deterioration and then to move to achieve this steady state condition. I would like to see increased energy needs met by the direct utilization of solar energy. There are, however, visionary scientists, and policy makers who will listen to them, who will grasp at any straw to keep man's exploitation of the earth forever growing. In my mind I can hear them planning to air condition part of the earth for man and for whatever then will supply his food, while the rest of the earth is allowed to radiate the excess heat by attaining a very high temperature. They will consider putting generating plants and factories on the moon and other planets and reducing the heat stress on earth by reflecting solar radiation back to space. This latter possibility brings us to consideration of some secondary effects of our expanding energy budget.

Side Effects of Energy Use

The combustion of fuel releases not only heat but also frequently smoke and various chemicals, the most important of which are water and carbon dioxide, into the environment.

Smoke and other particulate matter in the atmosphere scatters and absorbs incoming solar radiation, thus reducing the amount of energy absorbed by the surface. It has little effect on the outgoing longwave radiation, so the net effect is a tendency to cool the earth's surface. Several large volcanic eruptions within historic times have caused a year or so of abnormally low temperatures all over the earth. Perhaps the most striking case was the mighty eruption of Mt. Tomboro in the Lesser Sunda Islands in 1815. The following year, 1816, was the famous "year without a summer," during which snow fell in Boston every month. Actual measurements following the 1912 eruption of Mt. Katmai in the Aleutians showed a reduction of about 20% in the solar radiation reaching the earth.

Water vapor and carbon dioxide have an effect opposite to that of smoke; they are transparent to sunlight but absorb the longwave radiation from the earth and, by reradiating some of it back to earth, tend to raise the surface temperature. Man's combustion of fossil fuels has caused a measurable increase in the CO_2 content of the atmosphere, and now increasing numbers of jet airplanes are releasing great quantities of both water vapor and CO_2 at high altitudes. This would tend to raise the earth's temperature. However, a phenomenon of increasing frequency is the coalescence of the contrails of jet airplanes into banks of cirrus clouds which will reflect some of the incoming solar radiation back to space. Obviously, we cannot now be certain of the ultimate effects of the materials we are releasing to the atmosphere, but they certainly have the potential for changing climates.

On a more local scale, we are using prodigious quantities of water for cooling industrial plants, especially electrical generating plants, and this use is expected to increase at least as rapidly as our energy use grows[2]—perhaps more rapidly because nuclear plants waste more heat per kilowatt than plants burning fossil fuels. When heated water is discharged at a temperature above that of the air, we must expect an increase in the frequencies of mist and fog and, in winter, icing conditions.

Biological Effects of Thermal Pollution

The first and most obvious biological effect one thinks of is that bodies of water may become so hot that nothing can live in them. It is true that there are a few bacteria and blue-green algae that can grow in hot springs but even these are very unusual in water above 60 C. I know of only one case of a green alga living above 50 C—a species of *Protococcus* from Yellowstone Park. A few rotifers, nematodes, and protozoa have been found at above 50 C, and some, in a dried and encysted state, will survive much higher temperatures. In general, no higher organisms are to be expected actively living in water above about 35 C. I find it difficult to reconcile the data with the conclusion of Wurtz (1968, p. 139): "Water temperature would have to be increased to about 130 F (54.4 C) to destroy the microorganisms that are responsible for the self-purification capacity of a lake or stream." This statement is certainly incompatible with the recommendation of the National Technical Advisory Committee (1968) that: "All surface waters should be capable of supporting life forms of aesthetic value."

Another factor to be considered is the effect of temperature on the types of organisms present. Fish of any type are rare above 30 C. Diatoms, which are important members of aquatic food chains, decrease as temperature rises above 20 C, and they are gradually replaced by blue-green algae which are not important as food for animals, are often toxic, and are often the source of water blooms which kill the biota and make the water unfit for domestic use. Typically at about 30 C, diatoms and blue-greens are about equally represented and green algae exceed both. At 35 C, the diatoms are nearly gone, the greens are decreasing rapidly,

---
[2] Thermally more efficient electrical generating plants are considered possible: e.g., the "magnetohydrodynamic" generator (see Rosa and Hals, 1968).

and the blue-greens are assuming full dominance.

So far as animals are concerned, a body of water at a temperature of 30-35 C is essentially a biological desert. The green algae, which are well above their optimum temperature, can support a few types of cladoceran, amphiphod, and isopod crustaceans; bacteria and blue-green algae may abound; and mosquito larvae may do very well; rooted plants may grow in shallow regions; and a few crayfish, carp, goldfish, and catfish may endure. Largemouth bass can survive and grow at 32 C but they do not reproduce above about 24 C. A few other forms such as aquatic insects may inhabit the water as adults but may or may not be able to complete their life cycles there. Desirable game fishes such as Atlantic salmon, lake trout, northern pike, and walleyes require water below 10 C for reproduction.

Another effect of raising the temperature of water is to reduce the solubility of gases. The amount of oxygen dissolved in water in equilibrium with the atmosphere decreases by over 17% between 20 C and 30 C. At the same time, the need of organisms for oxygen increases. As a rule of thumb, metabolic rate approximately doubles for a 10 C increase in body temperature, although the effect is sometimes greater. Krogh (1914) found that the rate of development of frog eggs is about six times as great at 20 C as at 10 C. Similar effects have been noted in the rate of development of mosquitos.

Dissolved oxygen is often in critically short supply for aquatic organisms and increased temperature aggravates the situation. This may be partially compensated by more rapid diffusion of oxygen from the air and, during daylight hours, by photosynthesis. On the other hand, decay of organic matter and other oxidative processes such as the rusting of iron are more rapid at high temperatures. In polluted water the effect of biochemical oxygen demand (BOD) is more severe at high temperatures. The addition of heat to estuaries may be more critical than in bodies of freshwater because saltwater has a slightly lower specific heat and because oxygen is less soluble in saltwater.

In contrast to the situation with gases, salts become more soluble in water as the temperature increases. Chemical reactions become more rapid and increased evaporation may further increase the concentration of dissolved salts. At the same time the rate of exchange of substances between aquatic organisms and the medium increases. Toxins are likely to have greater effects, and parasites and diseases are more likely to break out and spread. In general, water at a temperature above the optimum places a strain on metabolic processes that may make adaptation to other environmental factors more difficult. For example, the Japanese oyster can tolerate a wider range of salinity in winter than in summer (Reid, 1961, p. 267).

In addition, if the water contains plant nutrients, objectionable growths of aquatic plants may be promoted by increased temperature—extreme cases leading to heavy mortality of fishes and other animals. It has been reported that polluted water from Lake Superior which does not support algal blooms there will do so if warmed to the temperature of Lake Erie.

Still other factors come into play when a body of water is thermally stratified. In a deep cold lake such as Cayuga where we are currently threatened with a huge nuclear generating plant, the lake water mixes, usually in May, so the entire lake is well oxygenated. As the surface warms, the lake stratifies with a level of light, warm water (the epilimnion) floating with no appreciable mixing on a mass of dense cold water (the hypolimnion) in which the lake trout, their food organisms, and other things are living and consuming oxygen. This continues until the lake mixes again, usually in November, by which time the oxygen supply in the hypolimnion is seriously depleted. The power company plans to pump 750 million gallons per day from the hypolimnion at a temperature averaging perhaps 6 C and to discharge it at the surface at a temperature of about 21 C. They plan this despite a recommendation of the National Technical Advisory Committee (1968, p. 33) that: "... water for cooling should not be pumped from the hypolimnion to be discharged to the same body of water." The effect of this addition of heat on the average temperature of the lake will be trivially small, but the biological consequences can be out of all proportion to the amount of heating.

The heat will delay fall cooling of the epilimnion and hasten spring warming, so that the length of time the lake is stratified each year will be increased. The water from the hypolimnion is rich in available plant nutrients, and by warming it and discharging it in the lighted zone, the amount of plant growth will be increased. This means more organic matter sinking into the hypolimnion and using up oxygen when it decays. The threat to the welfare of the lake is very real.

Finally, we should comment on fluctuating temperatures. Many organisms can adapt to somewhat higher or lower temperatures if they have time. The water in reservoirs behind hydroelectric dams often becomes thermally stratified, and when it is released at the base of the dam, a stretch of cold stream is produced which can support a cold water fauna even in warm regions. But when water is only discharged during peak electrical generating hours, the stream becomes subject to severe temperature fluctuations that will exclude many sensitive organisms. Similarly, if fishes or other organisms acclimate to the warm discharge from a factory or power plant and congregate near it, they will be subjected to temperature shocks when the plant is shut down for maintenance or refueling.

Conclusion

Man cannot go on increasing his use of thermal energy without causing degradation of his environment, and if he is persistent enough, he will destroy himself. There are other energy sources that could be used, but no source can support an indefinitely growing population. As with so many other things, it is man's irresponsible proliferation in numbers that is the real heart of the problem. There is some population size that the earth could support indefinitely without undergoing deterioration, but people do not even want to consider what that number might be. I suspect that it is substantially below the present world population. One would think that any rational creature riding a space ship would take care not to damage or destroy the ship, but perhaps the word "rational" does not describe man.

References

Brooks, C. E. P. 1949. *Climate Through the Ages.* Rev. ed. McGraw-Hill Book Co., New York.

Kardashev, N. S. 1964. Transmission of information by extraterrestrial civilizations. In: *Extraterrestrial Civilizations*, G. M. Tovmasyan (ed.). Trans. by National Aeronautics and Space Administration, Washington, D.C.

Krogh, A. 1914. On the influence of temperature on the rate of embryonic development. *Z. Allg. Physiol.*, **16**: 163-177.

National Technical Advisory Committee. 1968. Water quality criteria. Report of the National Technical Advisory Committee to the Secretary of the Interior. Federal Water Pollution Control Administration, Washington, D.C.

Putnam, P. C. 1953. *Energy in the Future*. D. Van Nostrand Co., New York.

Reid, G. K. 1961. *Ecology of Inland Waters and Estuaries*. Reinhold Publishing Co., New York.

Rosa, R. J., and F. A. Hals. 1968. In defense of MHD. *Ind. Res.*, June 1968: 68-72.

Waggoner, P. E. 1966. Weather modification and the living environment. In: *Future Environments of North America*, F. F. Darling and J. F. Milton (eds.). Natural History Press, Garden City, N.Y.

Wurtz, C. B. 1968. Thermal pollution: The effect of the problem. In: *Environmental Problems*, B. R. Wilson (ed.). J. B. Lippincott Co., Philadelphia.

Detergent Enzymes: Biodegradation and Environmental Acceptability

R. D. Swisher

Household laundry detergents have been developed wherein improved cleaning action has been achieved by incorporating enzymes into the formulations; use of such products may increase in the future (Ellwood, 1968). The present work was undertaken to assess the biodegradability of such an enzyme, since ready biodegradation would minimize likelihood of any effects following discharge into the general environment after use. Rapid destruction did indeed occur under bacterial action,

TABLE 1. Biodegradation in semicontinuous activated sludge test (SDA, 1965)

	Control (A)	Test (B)	Net (B-A)
Feed Composition, mg/liter			
Glucose	300	300	0
Nutrient broth	200	200	0
K_2HPO_4	130	130	0
Enzyme SG-2331	0	300 [a]	300
Organic carbon	200	300	100 [b]
24-Hour Effluent			
Protease, units/ml [c]	0	0	0
Organic C, mg/liter [d]	7.5	13.5	6
Organic C, % remaining	3.8		6

[a] Alkaline protease content of feed 90 units/ml.
[b] Determined directly as well as by difference. Organic carbon analyses by Beckman Carbon Analyzer, Model 915.
[c] Days 3, 4, 9. On days 1 and 2 apparent alkaline protease was 10 to 35 units/ml in A and B, with no significant difference between them.
[d] Days 7, 8. On day 1, B was 17 mg/liter, B-A 8.5 mg/liter.

TABLE 2. Enzyme (SG-2331) biodegradation in die-away tests

	Initial SG-2331 mg/liter	Alkaline protease, units/ml			
		Day 0	Day 1	Day 2	Day 3
Shake Culture					
Run 1: Control	0	0	1.3	0.2	
Run 1: SG-2331	30	10.5	10.1	2.0	
Run 2: Control	0	0	2.4	0.3	
Run 2: SG-2331	30	10.2	12.0	0.5	
River Water					
Control	0	0.3	0		
SG-2331	20	6.0	0.2	0	
Yeast extract, 300 mg/liter	0	0.2	0.1	1.5	0.3
Sewage					
Raw sewage: Control	0	0.4	0.3		
Raw sewage + SG-2331	20	6.7	0.3		
Secondary effluent: Control	0	0.1	0		
Secondary effluent + SG-2331	20	6.6	0		

as had been anticipated in view of the predominantly proteinaceous nature of enzymes, and in view of the earlier observation by Marion and Malaney (1963) that several enzymes were extensively oxidized by bacteria, oxygen uptakes ranging from 0.4 to 0.8 grams per gram of enzyme.

The enzyme used in these studies was a composite (SG-2331), representative of a commercial product, and appears to be typical of those used in detergents. It was produced from a strain of *Bacillus subtilis* and had alkaline and neutral protease activities of approximately 0.3 and 1.0 units/μg respectively. The actual protease content by weight was unknown, but was probably no more than about 10%, the remainder being mainly inert protein and polysaccharide. The unmodified term "enzyme" as used herein refers to such commercial-type products containing only a minor percentage of active enzyme.

Both alkaline and neutral protease activities

were determined throughout these experiments. In the interest of simplicity, only the alkaline protease data are presented in detail. The neutral is the more labile of the two under physical and chemical stress, and under the biological action its rate of disappearance equaled or exceeded that of the alkaline.

Detergent enzymes will enter the general environment via domestic sewage. Assuming an enzyme level of 1% in the detergent formulation (about 1/20 of the surfactant level), and assuming universal usage, the resulting enzyme concentration in domestic sewage would be around 1 mg/liter order of magnitude. If there is no destruction of enzyme activity during the laundering process (actually 50% or more may disappear), this corresponds to alkaline protease content of 0.3 units/ml in the sewage. The levels of SG-2331 used in these experiments were considerably higher (20, 30, or 300 mg/liter), dictated by the sensitivity limits of the enzyme assay methods.

Enzyme activity was determined by the Kunitz (1947) casein digestion method, with minor modifications. One unit of protease activity is defined herein as the amount liberating 0.5 µg of tyrosine (Folin-Ciocalteau reagent) during 10 min at 37 C, alkaline protease being determined at pH 10.2 and neutral at 7.0. Increased sensitivity was achieved by extending the digestion period to 100 min, during which time one unit of activity liberates approximately 5 µg of tyrosine.

Preliminary biodegradation experiments were made under the conditions of the Soap and Detergent Association (SDA) (1955) semicontinuous activated sludge test method, which involves aeration of a concentrated bacterial floc in a series of 24-hr cycles, the nutrient medium and tested material being replenished at the beginning of each cycle. A newer SDA feed composition shown in Table 1, and a routine enzyme assay method with 10-min digestion were used. Within the sensitivity limits of this assay method, complete destruction of the protease activity of SG-2331 (300 mg/liter feed concentration) occurred during the 24-hr cycle. No acclimation period was necessary. Not only the enzyme activity but also the entire substance of the SG-2331 was degraded: net residual organic carbon from the enzyme was 6% of that fed, compared to 4% remaining from the medium itself (Table 1). Thus, the bacteria of the activated sludge utilized both the active and the inert components of the enzyme as food to an extent comparable with glucose and nutrient broth.

The subsequent experiments were made with somewhat more realistic initial concentrations of 20-30 mg/liter and under milder conditions, in shake-cultures, river water, and sewage, using the more sensitive and reproducible 100-min enzyme assay procedure. Degradation of both alkaline and neutral protease occurred rapidly, within 1 to 2 days. Alkaline protease data are shown in Table 2.

The shake culture conditions corresponded to those of the SDA (1965) biodegradation procedure except that the flasks were not inoculated. The medium (500 ml containing mineral salts and 150 mg yeast extract) was incubated in 1-liter Erlenmeyer flasks loosely covered with aluminum foil on a rotary shaker (200 strokes/min, 2.5 cm diameter) at 24-26 C. The medium was not sterilized, and bacteria developed in it from the natural environmental population (Swisher, 1966). Disappearance of alkaline and neutral protease activity was substantially complete in 48 hr.

River water tests were made in half-filled, covered 2-liter wide-mouth jars, quiescent and in semidarkness. Degradation was essentially complete within 24 hr.

Sewage tests were made with raw sewage or treated effluent from a conventional activated sludge plant treating domestic sewage. After filtration through paper, 500 ml samples were shaken in 1-liter Erlenmeyers as in the shake culture test. Degradation of the added alkaline and neutral protease was complete within 24 hr.

The alkaline protease content, 0.4 units/ml given for the raw sewage control in Table 2 appeared to be characteristic—four samples taken at the plant over a 7-day period ranged from 0.4 to 0.7 units/ml; neutral protease ranged from 0.6 to 0.9 units/ml. This is comparable to the order of magnitude estimate of 0.3 units made above for alkaline protease from detergent enzymes if universal use occurs. These natural enzymes were degraded during the sewage treatment, since the effluents collected during the same period ranged from zero to 0.1 units/ml alkaline and zero to 0.2 neutral.

Formation and subsequent degradation of alkaline and neutral proteases was observed in the shake culture controls in both runs. Table 2 shows peaks of 1.3 and 2.4 units/ml at 24 hr for the alkaline; corresponding values for the neutral were 10.1 and 2.8. Addition of the shake culture nutrient (yeast extract, 300 mg/liter) to the river water gave a similar result under quiescent conditions: protease activities peaked at 48 hr, 1.5 and 2.1 units/ml of alkaline and neutral respectively, and were mostly gone at 72 hr.

In view of the ready biodegradability of the detergent enzyme, and the natural occurrence of similar enzymes at comparable levels, it seems unlikely that its use would cause environmental reactions either from direct action or from second order effects.

Acknowledgments

Thanks to P. L. McGowen and M. F. James for making the enzyme assays and to Drs. P. H. Hodson, S. G. Clark, and L. Keay for information and advice on the enzyme and assay methods.

References

Ellwood, P. 1968. Will enzymes trigger a detergent revolution? *Chem. Eng.*, **75** (20): 108-110.

Kunitz, M. 1947. Crystalline soybean trypsin inhibitor. II. General properties. *J. Gen. Physiol.*, **30**: 291-310.

Marion, C. V., and G. W. Malaney. 1963. The oxidation of aliphatic compounds by *Alcaligenes faecalis*. *J. Water Pollut. Contr. Fed.*, **35**: 1269-1284.

Soap and Detergent Association, Subcommittee on Biodegradation Test Methods. 1965. A procedure and standards for the determination of the biodegradability of alkyl benzene sulfonate and linear alkylate sulfonate. *J. Amer. Oil Chem. Soc.*, **42**: 986-993.

Swisher, R. D. 1966. Shake culture biodegradation of surfactants without inoculation. *Develop. Ind. Microbiol.*, **7**: 271-278.

R. D. SWISHER
Inorganic Chemicals Division
Monsanto Company
St. Louis, Missouri 63166

Oil Pollution

Damage observed in tropical communities along the Atlantic seaboard of Panama

Klaus Rützler and Wolfgang Sterrer

Observations on the effects of oil pollution on tropical marine habitats are reported. The pollution was caused by the wreckage of a tanker off the Atlantic entrance to the Panama Canal. Infralittoral communities such as coral reefs remained unaffected because no detergents have been used in eliminating the oil. Repopulation of intertidal rocks covered by dried tar took place 2 months after the incident. Greatest damage occurred on microfauna and intertidal organisms in sandy beaches and mangroves. (BioScience 20, no. 4, p. 222-224)

From 18 to 22 February 1969, approximately 2 months after the wreck of the oil tanker *Witwater* (13 December 1968), we had the opportunity to visit the site of the event to make a preliminary study of the effects of oil pollution on marine habitats.

The incident occurred approximately 2 nautical miles northeast of Galeta Island, Canal Zone, where the marine facility of the Smithsonian Tropical Research Institute is located (Fig. 1). The 3400-ton tanker S. S. *Witwater* ruptured on its way to the Atlantic entrance of the Panama Canal.

Fig. 2. Oil drifting toward the Galeta Island shore near the Marine Laboratory.

Fig. 1. Aerial view of the Smithsonian Tropical Research Institute Marine Laboratory, Galeta Island.

Close to 20 thousand barrels of diesel oil and Bunker C were released during and after the accident and driven toward the Galeta Island shoreline by the strong onshore seasonal winds (Fig. 2).

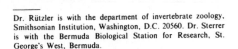

Dr. Rützler is with the department of invertebrate zoology, Smithsonian Institution, Washington, D.C. 20560. Dr. Sterrer is with the Bermuda Biological Station for Research, St. George's West, Bermuda.

By avoiding the use of detergents, which proved to be fatal to marine life during the well known Torrey Canyon wreckage (Scilly Isles) and by burning and pumping off the oil which had accumulated under the force of onshore winds in a small bay adjacent to the Galeta Laboratory, the Smithsonian Tropical Research Institute staff, in collaboration with members of the local military forces had considerably reduced the threat of severe damage to marine life.

Unfortunately, because of the very recent initiation of research on invertebrate groups at the Galeta Laboratory, information available about the composition of the nearby biota in the unpolluted state is incomplete and concentrates on vertebrate populations only. However, by comparing our observations with data from other better known areas in the tropical Atlantic, we obtained a good impression of the actual and potential damage to the flora and fauna of the various habitats caused by the oil spillage.

Rocky Shores

Since oil untreated with detergents is floating on the surface, it is the intertidal biota which suffer most from a spill. A great percentage of the affected shore consists of exposed rocky coast, mainly the Fort Randolph area and the causeway leading to the Galeta Laboratory. High winds caused a spray of mixed seawater and oil to cover trees and shrubs in the supralittoral zone to a height of 2 m above mean tide level; the oil has already killed many of these plants. Supralittoral spray pools and upper mesolittoral tide pools are still covered with a layer of oil up to 2 cm deep and are devoid of life (Fig. 3).

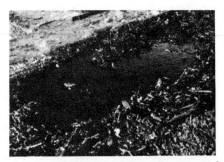

Fig. 3. Oil-covered supralittoral spray pool near Fort Randolph.

Thanks to the strong water agitation producing water-in-oil emulsions, the lower mesolittoral and the infralittoral zones were only temporarily affected by oil. The waves, as soon as they had discharged their oil, already started to erode it again. Also, the rapid action of the emergency crew, who by pumping and burning eliminated large quantities of oil (Fig. 4) before the tide had withdrawn, helped to lessen the effects in these zones. Nevertheless, damage to the gastropod and barnacle populations has to be assumed.

Wherever rocks and driftwood are fully exposed to the sun at low tide, the oil layer is slowly being reduced to a thin crust of tar and the substrates are being repopulated by the usual variety of intertidal organisms.

Fig. 4. Burning of oil which was trapped in a small bay next to the Galeta Marine Laboratory.

TABLE 1.

	Oil sand	Unaffected sand	
	Panama	Florida [a]	Morehead City, N.C.
Ciliata	45	-	32
Turbellaria	3 [b]	38	243
Nematoda	6	22	8.000
Archiannelida	-	100	-
Other Annelida	6	4	40
Copepoda	-	18	45
Others	-	2	430
Total	60	194	8.790

[a] Average of cores d-1, top 35 mm (Bush, 1966, p. 64)
[b] *Macrostomum* sp.

Coral Reefs

Although the water was very agitated and turbid during our visit, we were able to dive in shallow water. The reefs seemed to be the least affected communities of all. Shallow coral patches, consisting mainly of *Porites furcata, P. asteroides, Siderastrea radians, Millepora Complanata* (a hydrocoral), and associated organisms showed no ill effects at the time of the survey. This can be explained by the fact that these corals were subtidal and did not come into direct contact with the oil, which is mainly confined to the air-water interface. Further, due to high winds, water level at low tide was higher than usual; some of the corals observed would probably be partly exposed during very low tides, but they were not exposed or affected by oil during this period.

Sandy Beaches

Below a blackish oil-stained supralittoral zone, the sandy beaches of the affected area looked clean, at first glance. However, below a recently accumulated clean layer (2-5 mm), the sand was permeated completely with oil. In contrast to the rocky shore, sandy beaches as well as mangrove swamps have a rather horizontal than vertical extension (flat angle of exposure), and a large "internal" surface. They act as huge natural filters for the water masses brought in by waves and tides.

In the case of the sandy beach, still another phenomenon has to be considered, that of the subterranean backflow of water. With every wave breaking and running dead, a considerable amount of water is transported up the beach. Whereas most of the lower portion of this water body flows back on top of the sand, most of the upper portion filters vertically into the sand (Sterrer, 1965). Rough observations showed that, taking a wave period of 10 sec, and a water layer 1 mm thick as a basis, this means that 86,400 liters of water are filtered through 1 m of surfaces and per 24 hr. This considerable amount of water, then, together with the fresh groundwater, creates a continuous stream which—depending on local conditions—may continue for hundreds of meters underneath the sea bottom.

In the case of an oil spill, this means that the surface oil film brought ashore is deposited on top of the sand when the water soaks in and is pressed down by the following wave. The subterranean current, therefore, will mix more and more with oil which will invade biotopes far from the contaminated surface. On the positive side, the large surface area of the interstitial oil droplets will facilitate bacterial degradation.

Some quantitative samples of sand which we collected came from a small beach in the mangrove, approximately 2 km southeast from the Galeta Station. The sand (medium grain size, with a high amount of detritus), taken at midwater level, was heavily contaminated with oil to at least 30 cm sediment depth. An analysis[1] showed that 100 cm of sand (average wet weight = 121.9 g) held 31.8 g of water and 6.2 g oil. One fresh sample was brought back to the laboratory and studied alive, using the magnesium chloride technique described by Sterrer (1968).

Data for comparison with unaffected beach sediments of similar characters come from Bush (1966), who did her study on Florida beaches, and from unpublished samples collected in Morehead City, North Carolina, during a recent student excursion with Dr. R. Riedl (Chapel Hill). Whereas the Florida data seem to refer to a medium sheltered beach, the North Carolina material was collected on a very sheltered tidal sand flat. For better comparison, the number of specimens given in Table 1 have been calculated for 100 cm^3 of sediment.

The considerable difference in the nematode figures of the Florida and the North Carolina samples can be explained partly by the much more sheltered position of the latter collecting locality. This invariably means, on sandy beaches, that the absolute numbers as well as the dominance of nematodes increase. Another factor is that of the method—our samples from both Panama and North Carolina were washed with about 8 liters of magnesium chloride solution and then filtered through a plankton netting of 65 micron mesh width, whereas the Florida samples

---
[1] To determine the oil content of the samples, the sand was dried at 37 C over silica gel until no more weight loss occurred. Then it was repeatedly washed in benzene. The blackish stained benzene was then evaporated in large Petri dishes at 37 C and the residue weighed.

were treated with much less fluid and simply decanted.

Summarizing these few and very preliminary data, it seems that the oil inflow into the sand beach ecosystem results in a dramatic reduction of the meiofauna population. Obviously, animals with a relatively large (limbs) and inert (cuticula) body surface, such as crustaceans, are the first to disappear. The occurrence of comparatively large amounts of small ciliates might indicate the presence of oil-degrading bacteria on which they prey.

Mangroves

As could have been expected, the mangroves, being a predominantly intertidal community, suffered the most under the oil spillage. All the oil that could not settle along the exposed rocky shores was finally driven into the protected mangrove area (Fig. 5). The wide intertidal mud flats

Fig. 5. Intertidal stilt roots of *Rhizophora* in a badly affected mangrove area.

were all more or less thickly covered with oil. So was the surface of the tide channels distant from the open sea. Every footstep on the mud at low tide released large quantities of oil from the substrate. The pneumatophores of black mangrove trees (*Avicennia*) were all covered with a mixture of oil and mud. The stilt roots of the red mangrove (*Rhizophora*) had a thick layer of pure oil on their mesolittoral sections. It is still too early to judge damage on the fully grown trees, but it is to be expected that *Avicennia* in particular will suffer, since their pneumatophores are of vital importance for the ventilation of the remainder of the root system, which is buried in the anaerobic mud. The majority of young seedlings of *Rhizophora* were found to be killed, covered by oil (Fig. 6).

We did not obtain any data on the microfauna of the mud which could not be studied alive. A strong reduction of the

Fig. 6. Dead oil-covered seedlings of *Rhizophora*.

fiddler crab population (*Uca* sp.) could be observed in comparison to other mangrove areas. This is not surprising if we compare the oil content values of four mud samples taken in a mangrove swamp 1 km southeast of Galeta Marine Station, 500 m distant from the open water. The samples (5 cm surface layer) were selected from intertidal localities which appeared to be affected by a various degree (I-III):

Sample	Wet weight (g/100 cm³) (saturated with water)	Water (g/100 cm³)	Oil (g/100 cm³)
I	170.3	56.1	9.3
II	143.1	53.4	17.0
III	119.2	50.9	21.4

The characteristic intertidal algal community "*bostrychietum*" on the *Rhizophora* stilt roots and its inhabiting microfauna were practically eliminated in all oil exposed areas (Fig. 7); as were the sedentary animals of this zone, such as oysters (*Crassostrea* sp., Fig. 8), mussels (*Brachidontes* sp.), barnacles (*Balanus* sp.), sponges, tunicates, and bryozoans (compare: Rützler, 1969).

The infralittoral horizon in the mangroves consists of numerous tide channels and lagoons. There the wickerwork of *Rhizophora* stilt roots provides protection for many juvenile and adult fishes and various crustaceans, many of which are of commercial importance. All of these are directly or indirectly (via the food

Fig. 7. Oil-smeared *bostrychietum* on *Rhizophora* roots.

Fig. 8. Dead oil-covered oysters on *Rhizophora* stilt roots.

chain) endangered by oil pollution. Obviously affected are those species which, for respiration, or for obtaining food, have to penetrate the air-water interface. We have observed dead and dying young sea turtles (*Caretta* sp.) on mangrove beaches; mass mortality of seabirds had been reported by the Torrey Canyon Report; a number of oil-smeared herons and one dying commorant were observed by a staff member of the Smithsonian Tropical Research Institute, near Galeta.

Conclusions

This fragmentary survey taught us the importance of immediate research on the effects of oil pollution in the sea, so that we may be ready for possible future oil spills; with increasing use of larger tankers, wrecks in the future could be catastrophic.

Even some of the long-term effects of the fortunately moderate accident off Galeta have yet to be investigated. We should take advantage of this warning and concentrate on research at places, such as Galeta, where the conditions of the natural environment can be studied, as well as the effect upon those conditions of experimentally introduced oil spills. Physical and chemical as well as biological phenomena must be studied in this connection. STRI has already proposed one such effort.

Acknowledgment

We are grateful to Dr. P. Glynn, Mr. E. Kohn, and Dr. I. Rubinoff of the Smithsonian Tropical Research Institute for their hospitality and valuable advice. Dr. Rubinoff provided Figures 1, 2, and 4.

References

Bush, L. 1966. Distribution of sand fauna in beaches at Miami, Florida. *Bull. Mar. Sci.*, **16**: 58-75.

Rützler, K. 1969. The Mangrove Community, Aspects of its Structure, Faunistics and Ecology. Proc. International Symposium on Coastal Lagoons, Mexico City, 1967 (in press).

Sterrer, W. 1965. Zur Oekologie der Turbellarien eines suedfinnischen Sandstrandes. *Botanica Gothoburgensia*, **3**: 211-219.

———. 1968. Beitraege zur Kenntnis der Gnathostomulida. I. Anatomie und Morphologie des Genus *Pterognathia* Sterrer. *Arch. Zool.*, **22**: 1-125.

Toward Safer Use of Pesticides

Sheila A. Moats and William A. Moats

About 300 pesticide chemicals with very diverse properties were reported to be in use in 1966. Some are inherently more hazardous to fish and wildlife and to man than others. Highly toxic organophosphate insecticides are very hazardous to agricultural workers since they allow little margin for carelessness or misuse. Human illnesses and deaths from pesticides could be reduced by substituting less toxic compounds where they will serve the purpose. Reported injurious effects of pesticides to fish and wildlife result mainly from a limited number of organochlorine insecticides, especially DDT and related compounds. Heavy applications of DDT to control Dutch Elm disease have been shown to cause severe losses of birds in the area of application. Use of the less toxic pesticide, methoxychlor, or sanitation, to control the disease, can reduce or eliminate the hazards to birds. Environmental contamination by organochlorine insecticides is of concern because these compounds can be, and have been, biologically concentrated to injurious levels in a number of cases. Levels harmless to adults can adversely affect hatchability of eggs of birds and fish. There is considerable evidence linking environmental contamination with DDT and derivatives to reproductive failures in certain species of hawks and eagles, resulting in sharp declines in numbers. In many instances, alternative pesticides are available, which may be used with substantially less hazard to man and his environment. (BioScience 20, no. 8, p. 459-464)

The widespread use of new synthetic organic pesticides has introduced an unprecedented array of chemicals in the environment. According to Mitchell (1966), in 1966, approximately 300 organic pesticide chemicals were in general use—as insecticides, miticides, herbicides, fungicides, and for other miscellaneous purposes—in 10,000 different formulations. These chemicals were developed for control of specific pests and with so great an array of compounds, it is impossible to fully evaluate their effects on all possible nontarget organisms. It may be expected that in so diverse a group of chemicals, the undesirable biological side effects will be quite variable. Information is gradually accumulating which enables us to appreciate the nature of these side effects.

The general public first became aware of the potential hazards of pesticides in 1962 from Rachel Carson's book *Silent Spring* (1962). This book aroused a storm of criticism but did bring public awareness of a need for more study of the possible hazards of these compounds.

Dr. Sheila A. Moats is associate professor of biology at the Federal City College, Washington, D.C. Dr. William A. Moats, Beltsville, Md., is a biochemist and author of a number of papers on pesticide residue analysis.

Pesticides are generally recognized to be indispensable in modern agriculture although the benefits may sometimes be overestimated. The spectacular successes of insecticides in controlling diseases spread by insect vectors are also well established. Despite progress in biological control, pesticides are likely to remain our first line of defense against various types of pests for some time to come. However, Headley and Lewis (1967) point out that all too many discussions of the use of pesticides are written from a defensive point of view, and they point out the need for a more sophisticated economic approach to the use of pesticides.

Some Costs and Benefits of Pesticides

Benefits of pesticides may fall into several categories: increases in crop production resulting from applications of insecticides, herbicides, fungicides, etc.; preservation of other materials from attack by insects, fungi, etc.; control of nuisance-type insects; or reduction of deaths and illness from diseases through control of insect vectors.

Costs include the cost of the pesticide itself plus the cost of application, deleterious effects to human and nontarget plant and animal life, costs of monitoring for residues, and losses from destruction of foods which contain levels of residues considered to be excessive. Some costs and benefits may be estimated economically while others may be difficult to evaluate in monetary terms but are perhaps no less important. It is, for example, difficult to set a monetary value on human lives saved or lost through pesticide use or even on the positive or negative effects on human health.

Wild mammals, fish, birds, and other wild creatures are attractive esthetically and are perhaps of more economic importance than is generally realized. Apart from purely esthetic considerations, annual expenditures of hunters and fishermen are estimated by the U.S. Fish and Wildlife Service to be of the order of four billion dollars (Headley and Lewis, 1967). To put this figure in proper perspective, it approximately equals the farm value of all the corn produced in the United States and is nearly twice the farm value of cotton (USDA, 1967a). We can add to this the substantial expenditures of birdwatchers, hikers, and other nature-centered activities of millions of people. In addition, wild game and fish may provide an important source of quality protein food. Therefore, it is evident that fish and wildlife must be given adequate consideration in any cost/benefit analysis of pesticide use.

Injurious Effects

Examination of the literature indicates that injurious effects of pesticides are confined mainly to a limited number of compounds. Environmental contamination and most of the injurious effects of fish and wildlife reported result from the use or misuse of a small number of organochlorine insecticides. Although the older arsenicals apparently still cause more deaths annually (Hayes and Pirkle, 1966), most human illnesses and deaths from the newer synthetic pesticides are attributed to parathion and other highly toxic organophosphate insecticides through accidents or misuse, suicides, etc. The organochlorine insecticides and parathion also happen to be the cheapest to use and are, therefore, used extensively (Mitchell, 1966). It may appear that more injurious effects are found with these compounds simply because they are used so widely. However, they have certain properties which make them inherently more hazardous than many other insecticides. Parathion, which is highly toxic to warm-blooded animals including man (Mitchell, 1966; USDA, 1967b), is a significant hazard to agricultural workers. It breaks down rapidly, however, and seems to present little hazard in the environment or in foods. Many organochlorine insecticides are quite persistent and break down slowly. Their main hazard is not that they are persistent, as often stated, but that they are biologically concentrated. Were it not for this concentration, trace amounts in the environment would be of little concern. With the potential for concentration, very low levels may build up to injurious levels in certain organisms.

Effects on Human Health

Data on human illnesses and deaths from pesticides are not systematically tabulated for most areas. Hayes and Pirkle (1966) showed 111 accidental deaths attributed to pesticides in 1961 in the United States. This figure may be compared with 323 accidental deaths in one year from barbiturate poisoning (Headley and Lewis, 1967). More than half of the deaths attributed to pesticides were caused by pre-DDT pesticides, mainly arsenicals, and a substantial portion of the deaths occurred among small children. Dade County, Florida, reported 68 deaths from pesticides in the period 1956-65, 40 of which were caused by organophosphate insecticides (Davis et al., 1966). Some of these deaths were homicides or suicides, and the rest resulted mainly from accidents or misuse. For 1960-63, California reported 800-1100 cases annually of occupational illnesses from pesticide use among agricultural workers. These data do not cover the one-third of agricultural workers who are self-employed and do not include illnesses caused by pesticides among the general public.

In California, the rate of occupational illness among agricultural workers is reported to be higher than for any other industry (West and Milby, 1965). Many factors other than pesticides are involved; however, the incidence of workman's compensation awards for conditions resulting from inhalation, absorption, and swallowing of pesticides was three times higher than for all industry in 1961 (Kay, 1965). A quotation from Barnes (1966) gives an idea of the situation in some other parts of the world.

> "A recent report to WHO indicates what is happening in one small country of Central America where parathion and methyl parathion are being used on cotton. To quote an excerpt:
> 'The Departmental Hospitals at U- and S- each see up to 300 serious intoxications a month during the cotton growth season (about 6 months) with 2-3 deaths per month. Still more cases are treated in the field by foremen, administrators, and friends....'"

While most accidental deaths and illnesses result from carelessness and misuse, West (1966) cites about 400 illnesses among fruit pickers reported to have been caused by parathion residues on the leaves. In this case, the parathion was presumably applied according to accepted procedures.

From the California data, we may hazard a guess at the incidence of pesticide poisoning cases nationwide which required medical treatment. California uses about 10% of the total pesticides used in the United States (Andrilenas et al., 1969). There are 15,000 cases of poisoning among agricultural workers reported annually in the nation. Since the California data do not cover the one-third self-employed agricultural workers, we can assume that rather than the previously stated 800-1100 figure, there are about 1500 cases of poisoning that occur annually in California. About half the reported accidental deaths attributed to pesticides are among nonagricultural workers (Headley and Lewis, 1967). Therefore, it is reasonable to suppose that as many cases of serious poisoning occur among nonagricultural workers, giving us a total of 30,000 cases nationwide sufficiently serious to require medical attention. Considering that California makes a more active effort to control pesticide use than many other areas (Rudd, 1964; West and Milby, 1965), the figure of 30,000 illnesses requiring medical treatment annually appears conservative. West and Milby (1965) describe a special study made in Dade County, Florida, which showed 13 deaths from pesticides in 1963 alone—eight accidental and five suicides. If the special study had not been made, eight of the 13 deaths would have been attributed to causes other than pesticides. These figures, if representative for the nation, suggest that the number of 111 accidental deaths annually reported by Hayes (Hayes and Pirkle, 1966) for 1961 may be low by a factor of 2-3. It is difficult to reconcile a report of eight accidental deaths in one Florida county in one year with a nationwide total of only 111 annually.

It has been reported that acute or chronic poisoning by organophosphate insecticides sometimes results in long-term neurologic disorders (Faerman, 1967; West, 1968). West and Milby (1965) report that agricultural workers heavily exposed to pesticides are considered poor risks around machinery even if they do not show obvious symptoms of poisoning.

While few, if any, illnesses or deaths occur among users of pesticides who follow instructions and wear proper protective clothing, there is little margin for error with the highly toxic organophosphate insecticides. These compounds are frequently handled by people who have no appreciation of their toxicity (West and Milby, 1965). In selecting pesticides, one must consider their safety under actual, as well as ideal, conditions of use. Allowance should be made for the possibility that they might be handled carelessly or misused.

Human Loads of Chlorinated Pesticides

Great emphasis has been put on pesticide residues in foods with the results that they are generally relatively low (Hayes, 1966). Loads of pesticide residues in human body fat have been estimated to be 12 ppm for DDT and related compounds in the United States (West, 1966) and about 3 ppm of these compounds in Great Britain (Abbott et al., 1968) and Belgium (Maes and Heyndrickx, 1966). Small amounts of several other organochlorine pesticides were also found. Average levels of DDT and related compounds have remained constant in the United States since 1950 (Hayes, 1966) and have decreased slightly in recent years in Great Britain (Abbott et al., 1968). A summary by Robinson (1969) of human residue loads shows the wide variations found in different individuals. Human residue loads are generally far below levels known to cause intoxication (Hunter, 1968). Data from Hayes et al. (1958) indicate that levels of DDT and related compounds were only about 25% of the average in strict vegetarians, indicating that animal products are the main source of residues in human body fat. The only pesticides occurring in significant amounts in animal products are organochlorine compounds.

There is no direct evidence that present levels of pesticide residues in the human diet or in human body fat are harmful. However, it has been found that comparatively low levels of DDT and dieldrin in the diet induce an increase of microsomal enzymes (Durham, 1968; Kupfer, 1968) in the liver, which affect drug and steroid metabolism. These effects have been noted at levels of as little as 2.5 ppm DDT in the diets of rats (Fillette, 1968). The increases in microsomal enzymes are produced by many other chemicals besides pesticides and are evidently detoxification mechanisms. (The main significance in man appears to be that an increased rate of drug metabolism can reduce human response to the drugs.)

The o,p isomer of DDD has been found to depress the functioning of the adrenal cortex in a number of species, including man. The compound has been used successfully to treat Cushing's syndrome, a condition characterized by oversecretion of adrenal cortical hormones (Kupfer, 1968). The pesticide o,p-DDD was found to block the action of vitamin D_3 in mobilizing calcium in rachitic chicks (Sallis and Holdsworth, 1962) which may be significant in light of recent evidence, discussed further on, of disturbance of calcium metabolism in some species of raptorial birds. Reports that o,p-DDT has estrogenic activity were recently confirmed by Bitman et al. (1968). Technical DDT contains about 20% of the o,p isomer (Gunther and Jeppson, 1960). The pesticide p,p-DDT can be converted to p,p-DDD in rat livers (Datta et al., 1964), and it is likely that the o,p isomer can undergo similar conversion. Unfortunately, most published residue analysis for DDT, DDD, and DDE are based on the p,p isomers so there is no way of assessing the practical importance of residues of the o,p isomers in biological systems.

Effects of Fish and Wildlife

Most reported injurious effects of pesticides to fish and wildlife have involved a small group of organochlorine insecticides. In considering hazards to wildlife, we may distinguish the effects of wildlife in areas directly treated with pesticides from those resulting from general environmental contamination. Direct treatment may be disastrous to wildlife in the area treated but does not endanger wildlife generally. If treatment is discontinued, and the area treated is not too large, complete recovery may be expected in a few years. Where more extensive areas are treated, as was the case in the fire and control program, recovery may be slow (Rudd, 1964). Effects on wildlife from general environmental contamination, on the other hand, could be more serious since much larger areas are involved. The very existence of affected species may be threatened, and the contamination cannot be controlled.

The classic example of the effects of treating an aquatic ecosystem with an organochlorine insecticide (DDD) was described by Hunt and Bishoff (1960) and has been summarized by a number of authors. Clear Lake, California, was treated with low levels of this pesticide to control midges. Extensive concentration of this pesticide occurred in the food chain resulting in levels of up to 2500 ppm in the visceral fat of fish and extensive poisoning of grebes (*Aechmophorus occidentalis*) inhabiting the lake. Concentrations of pesticide in the edible flesh of some fish approached 200 ppm, far above legal tolerances. There have been numerous reports of severe poisoning of birds, particularly robins (*Turdus migratorius*), in areas heavily treated with DDT to control Dutch elm disease. These are summarized by Wurster et al. (1965). Poisoning is thought to result mainly from accumulations in insects and worms on which the birds feed. On the other hand, purple grackles (*Quiscalus quiscula*) and red wing blackbirds (*Agelaius phoeniceus*), both of which may be agricultural pests, were reported to be unaffected in areas heavily treated with DDT (Walley et al., 1966).

Carson (1962) and Rudd (1964) have summarized the adverse affects on wildlife and domestic animals of heavy pesticide treatments used in fire ant and Japanese beetle control programs.

Organochlorine insecticides are quite stable in the environment and are readily transported in air, on dust particles (Cohen and Pinkerton, 1966), or in water, either dissolved or adsorbed on particles of suspended organic matter (Keith, 1966). Their dissemination in the environment is, therefore, widespread and uncontrollable. Living organisms have a tremendous capacity to concentrate organochlorine pesticides, especially in food chains where successive concentration occurs as small organisms are consumed by larger ones. Therefore, low concentrations in the environment cannot be assumed to be harmless. Hunt (1966) cites a number of examples of such concentration in natural systems to levels injurious to organisms at the top of food chains, mainly fish-eating birds. In the Clear Lake example (Hunt, 1966) mentioned previously, concentrations in the fat of fish-eating birds were 100,000 times those applied to lake water; a number of instances have been noted where organochlorine insecticides have been concentrated several thousandfold. The dynamics of concentration of organochlorine pesticides have been discussed by Robinson (1967) and involve an equilibrium between intake, metabolism, and excretion. The equilibrium level attained at a given intake of pesticide depends on the physiology of the particular organism involved. Aquatic ecosystems are particularly susceptible since the food chains are more complex than in terrestrial systems and the opportunity for biological concentration of pesticides is, therefore, greater. Wide dissemination of organochlorine insecticides—particularly DDT and metabolites—and dieldrin is shown by findings of substantial concentrations in seals and porpoises in the North Atlantic (Holden and Marsden, 1967), in seabirds off California (Risebrough et al., 1967), and even in penguins and seals in the Antarctic (Sladen et al., 1966).

Robinson et al. (1967) observed that only HEOD (dieldrin) and p,p-DDE were found in significant amounts in marine organisms. They observed seasonal fluctuations in storage of these compounds, showing that results of single analyses from one season must be interpreted cautiously.

Consideration of DDT to near-toxic levels has been reported in a Long Island salt marsh, and the biota of this area might be significantly affected (Woodwell et al., 1967). A number of fish kills have been observed resulting from pesticide runoff into streams (Rudd, 1964). Ferguson (1967) has found that some organisms such as mosquito fish (*Gambusia affinis*) have become resistant to endrin and can accumulate sufficient pesticide in their bodies to poison predators feeding on them. He notes that large-mouth bass have disappeared in areas where such resistant fish occur, indicating that the effect may be ecologically significant. Resistance to endrin has also been found in sunfish (*Lepomis* sp.), thus presumably presenting a potential hazard to anyone unfortunate enough to eat one.

Effects on Reproduction

Where animals are not directly poisoned, reproduction may be affected; this can be as serious as direct poisoning. DeWitt (1955) found that the viability of pheasant (*Phasianus colchicus*) and quail (*Colinus virginianus*) eggs was affected by levels of DDT and dieldrin in the diets which did not harm the adults. Where eggs hatched, chicks frequently died a few days after hatching. Environmental levels of organochlorine insecticides appear to be high enough to have affected the reproduction of some species of birds. The evidence is summarized by Wurster and Wingate (1968) and by Hickey and Anderson (1968). Declines in reproductive rates have been noted in gulls (*Larus argentatus*), the Bermuda petrel (*Pterodroma cahow*), and several species of hawks and eagles. The existence of some species of hawks and eagles including the osprey (*Pandion haliaetus*) and the bald eagle (*Haliaeetus leucocephalus*) appears threatened by these reproductive failures. One species, the peregrine falcon (*Falco peregrinus*) has already been exterminated over a large portion of its former range (Hickey and Anderson, 1968). The declines began coincident with large scale use of DDT. A concomitant decrease in eggshell thickness occurred concurrently with this decline indicating derangement of mineral metabolism (Hickey and Anderson, 1968; Ratcliffe, 1967). DDT and other organochlorine insecticides are known to affect steroid metabolism in various species (Kupfer, 1968) and are thus prime suspects (Hickey and Anderson, 1968). Residues of p,p-DDE in eggs of peregrine falcons and sparrow hawks (*Accipiter nisus*) in Great Britain, both species which are declining, were found by Walker et al. (1967) to be higher than for most other species sampled. These results may reflect differences in metabolism or greater exposure from food sources. Declines in reproduction with ospreys were shown to be proportional to pesticide levels in the eggs (Ames, 1966). The evidence, therefore, points to chlorinated hydrocarbon pesticides as prime factors in the recent declines of these species. The species affected are all predators at the top of food chains; thus, the opportunity for exposure to concentrations of residues magnified biologically is at a maximum. It has been noted recently that polychlorinated biphenyls, industrial chemicals, are also widely distributed in the environment and also induce proliferation of microsomal enzymes. Risebrough et al. (1968) have found that concentrations of polychlorinated biphenyls in living organisms are somewhat lower than p,p-DDE, but these chemicals may also be involved in the declines noted in raptorial birds.

The viability of fish eggs was also found to be affected adversely by DDT. Losses of newly hatched fry of lake trout in a New York State hatchery were traced to DDT residues in the eggs (Burdick et al., 1967). The effects on reproduction of wild fish populations have not been reported but could be significant in some cases.

Crayfish are a by-product of rice growing and sometimes are more valuable than the rice. However, residues of aldrin and dieldrin in the crayfish (*Procamberus clarkii*) were found to be higher than could be permitted in interstate commerce. Residues appeared to result from environmental contamination rather than treatment of rice seeds used in the fields studied (Hendrick et al., 1966).

Some Pros and Cons of DDT

The spectacular results obtained by using DDT to control insect disease vectors have been well documented (Jukes, 1963) and one can scarcely criticize this use of DDT. At the time these programs were undertaken, there were few alternative pesticides and there was no other feasible method of controlling these diseases so rapidly and effectively. However, because DDT was the best method of controlling insect-borne disease in the 1940's, does not necessarily imply that it is still the method of choice. The diseases have been reduced greatly but not eradicated; resistance of insects to DDT is widespread, and many alternative pesticides are available today.

Use of heavy DDT sprays to control Dutch elm disease is an example of benefits that are purely esthetic. Elm trees provide neither food nor fiber, and human illness is not involved. Elm trees are merely pretty to look at. The value of birds is also mainly esthetic, though it may be argued that they eat insects. It is difficult to justify saving elm trees by a method known to kill large numbers of birds (Wurster et al., 1965). The use of an alternative pesticide, methoxychlor, has been reported to be as effective as DDT in controlling Dutch elm disease with much less hazard to birds and other wildlife (Whitten and Swingle, 1964). Prompt destruction of dead or dying elm trees has been reported by Mathysse (1959) to be effective in control of the disease. Whitten and Swingle (1964), however, conclude that this method was of little value. Consideration of benefits vs. costs would indicate to us that the use of DDT was unjustified in this case because of its injurious effects. Use of methoxychlor is more expensive but is largely free of injurious effects. Sanitation, if successful, would be cheaper than a method using pesticides because dead or dying trees would have to be removed sooner or later anyhow. Of course, sanitation is completely free of injurious side effects.

Some Suggestions for Action

We have considered, up to this point, some of the most serious examples of injurious effects known to be caused by pesticides, or for which there is strong evidence that they are caused by pesticides. Can these injurious effects be reduced or eliminated without losing the benefits of pesticides? It is likely that they can be, in many cases, simply by selecting alternative pesticides. The alternative pesticides are likely to be more expensive when only the cost of the pesticide material is considered and they may be slightly less effective against given pests. Realistic consideration of the true costs of any pesticide must consider any injurious effects, accidental or not, as part of the true cost of use of a pesticide. With organochlorine insecticides, costs of monitoring for environmental residues must be considered. With highly toxic pesticides, the costs of medical treatment and time lost through illness must be considered. A realistic cost/benefit assessment using this approach will tend to favor the use of nonpersistent pesticides of low toxicity to man. We have already discussed the substitution of methoxychlor for DDT for control of Dutch elm disease. The *USDA Guide to the Use of Insecticides* (1967b) lists several alternative insecticides for most applications, showing that it is frequently feasible, on the basis of present knowledge, to substitute pesticides of low persistence and toxicity. The dairy industry has successfully eliminated the use of persistent organochlorine insecticides in production of feeds and forages and in the control of insects affecting dairy cattle. The USDA (1965) has also eliminated broadcast applications of organochlorine insecticides from many of its pest control programs; for example, low-volume malathion sprays have been substituted for dieldrin for grasshopper control. In many cases, it has proved necessary to find alternatives to organochlorine insecticides because target insects have become resistant to them.

Ample information is now available on problems of pesticide usage to provide a basis for specific action to reduce injurious effects. Some specific suggestions for action are:

1) Restrict use of DDT and dieldrin, which are the most serious environmental contaminants. Regulatory agencies should restrict their use of these compounds to situations where such usage is of significant benefit to human welfare and where it can be shown that no alternative method of insect control is feasible. Cautious use in public health programs might, for example, be justified. Protection of purely ornamental plants would not be essential. Such action would significantly reduce environmental contamination without totally precluding the use of these compounds where such usage is absolutely essential. Similar action should be considered for other chlorinated pesticides. Steps to implement this approach have already been taken by the U.S. Department of Agriculture and several states.

2) Restrict use of highly toxic organophosphorus insecticides, such as parathion, until it is demonstrated that they can be used reasonably safely under actual conditions with some margin for error.

3) The Public Health Service should systematically collect data on pesticide poisonings which are clearly a significant public health problem. Such data would provide a more accurate appraisal of the magnitude of the problem and provide a rational basis for corrective action.

4) While steps are taken to reduce pesticide hazards, research on biological control should, at the same time, be accelerated. The question arises as to whether we can safely and effectively eliminate the use of pesticides. At the same time, we must substitute biological control cautiously, making sure at every stage that the balance of nature is undisturbed.

Analysis of costs and benefits of pesticide use is very complex, and judgments must be continually modified on the basis of experience and new research findings. Where evidence is presented, indicating that pesticides may be producing injurious effects, it seems reasonable that the burden of proof of safety should lie with the user or those who advocate the use of the pesticide in question. We believe that public policy can best be developed through free and open discussion of controversial issues from various points of view and hope that this paper will make a contribution toward the safer use of pesticides.

Summary

About 300 pesticide chemicals with very diverse properties were reported to be in use in 1966. Some are inherently more hazardous to fish and wildlife and to man than others. Highly toxic organophosphate insecticides are very hazardous to agricultural workers since they allow little margin for carelessness or misuse. Human illnesses and deaths from pesticides could be reduced by substituting less toxic compounds where they will serve the purpose. Reported injurious effects of pesticides to fish and wildlife result mainly from a limited number of organochlorine insecticides, especially DDT and related compounds. Heavy applications of DDT to control Dutch elm disease have been shown to cause severe losses of birds in the area of application. Use of the less toxic pesticide, methoxychlor, or sanitation, to control the disease, can reduce or eliminate the hazards to birds. Environmental contamination by organochlorine insecticides is of concern because these compounds can be, and have been, biologically concentrated to injurious levels in a number of cases. Levels harmless to adults can adversely affect hatchability of eggs of birds and fish. There is considerable evidence linking environmental contamination with DDT and derivatives to reproductive failures in certain species of hawks and eagles, resulting in sharp declines in numbers. In many instances, alternative pesticides are available, which may be used with substantially less hazard to man and his environment.

Note

Chemical names of pesticides mentioned in this paper are: p,p-DDD—1,1 dichloro-2,2-bis (p-chlorophenyl) ethane, o,p-DDD—1, 1 dichloro-2-(o-chlorophenyl)-2-(p-chlorophenyl) ethane, p,p-DDE—1,1-dichloro-2,2-bis (p-chlorophenyl) ethylene, p,p-DDT—1,1, 1-trichloro-2,2-bis (p-chlorophenyl) ethane, o,p-DDT—1,1,1-trichloro-2 (o-chlorophenyl)-2-(p-chlorophenyl) ethane, Dieldrin—1,2,3,4,10,10-hexachloro-*exo*-6,7-epoxy-1,4,4a,5,6,7,8,8a-octahydro-1,4,5,8-*endo*, *exo*-dimethano-naphthalene, Endrin—1,2,3,4,10,10-hexachloro-*exo*-6,7,-epoxy-1,4,4a,5,6,7,8,8a-octahydro-1,4,5,8-*endo*, *endo*-dimethanonaphthalene, parathion—O,O-diethyl O-p-nitrophenyl phosphorothioate, methyl parathion—0-0-dimethyl, 0-p-nitrophenyl phosphorothioate, methoxy-chlor—1,1,1 trichloro-2,2-bis (p-methoxyphenyl) ethane.

References

Abbott, D. C., R. Goulding, and J. O'G. Tatton. 1968. Organochlorine pesticide residues in human fat in Great Britain. *Brit. Med. J.,* **3** (5611): 146-149.

Ames, P. L. 1966. DDT residues in the eggs of the osprey in the Northeastern United States and their relation to nesting success. *J. Appl. Ecol.,* **3** (suppl.): 87-97.

Andrilenas, P., T. Eichers, and A. Fox. 1969. Farmers' expenditures for pesticides in 1964. Agr. Econs., Rept. No. 106. USDA, Econ. Res. Serv., 10 p.

Barnes, J. M. 1966. Human health and pest control. In: *Scientific Aspects of Pest Control*, Nat. Acad. Sci.-Nat. Res. Council, Publ. 1402, Washington, D.C., p. 435-452.

Bitman, J., H. C. Cecil, S. J. Harris, and G. F. Fries. 1968. Estrogenic activity of o,p-DDT in the mammalian uterus and avian oviduct. *Science,* **162**: 371-372.

Burdick, G. E., E. J. Harris, H. J. Dean, T. M. Walker, Jack Skea, and David Colby. 1964. The accumulations of DDT in lake trout and the effect on reproduction. *Trans. Amer. Fish Soc.,* **93**: 127-136.

Carson, Rachel. 1962. *Silent Spring.* Houghton-Mifflin Co., Boston, Mass.

Cohen, J. M., and C. Pinkerton. 1966. Widespread translocation of pesticides by air transport and rainout. In: "Organic Pesticides in the Environment." *Advan. Chem. Ser.,* **60**: 163-176.

Datta, P. R., E. P. Laug, and A. K. Klein. 1964. Conversion of p,p-DDT to p,p-DDD in the liver of the rat. *Science,* **145**: 1052-1053.

Davies, J. E., J. H. Davis, D. E. Frazier, J. B. Mann, and J. O. Welke. 1966. Urinary p-nitrophenol concentrations in acute and chronic parathion exposures. In: "Organic Pesticides in the Environment." *Advan. Chem. Ser.,* **60**: 67-68.

DeWitt, J. B. 1955. Effects of chlorinated hydrocarbon insecticides upon quail and pheasants. *J. Agr. Food Chem.,* **3**: 672-676.

Durham, W. F. 1968. The interaction of pesticides with other factors. *Residue Rev.,* **18**: 21-103.

Faerman, L. S. 1967. Late sequelae of acute poisoning with organic phosphorus insecticides. *Gig. Tr. Prof. Zobol.,* **11**: 39-41. (From *Chem. Abstr.,* **67**: 4003, 1967).

Ferguson, D. E. 1967. *The Ecological Consequences of Pesticide Resistance in Fishes.* Trans. 32nd N. Amer. Wildlife and Nat. Res. Conf., March 13-15, 1967. Wildlife Manag. Inst., Washington, D.C., p. 103-107.

Gillette, J. W. 1968. No effect level of DDT in induction of microsomal epoxidation. *J. Agr. Food Chem.,* **16**: 295-297.

Gunther, F. A., and L. R. Jeppson. 1960. *Modern Insecticides and World Food Production,* John Wiley & Sons, Inc., New York, 284 p.

Hayes, W. J., Jr. 1966. Monitoring food and people for pesticide content. In: *Scientific Aspects of Pest Control.* Nat. Acad. Sci-Nat. Res. Council, Publ. No. 1402, Washington, D.C., p. 314-342.

Hayes, W. J., Jr., and C. I. Pirkle. 1966. Mortality from pesticides in 1961. *Arch. Environ. Health,* **12**: 43-55.

Hayes, W. J., Jr., G. E. Quinby, K. C. Walker, J. W. Elliott, and W. M. Upholt. 1958. Storage of DDT and DDE in people with various degrees of exposure to DDT. *Arch. Ind. Health,* **18**: 398-406.

Headley, J. C., and T. N. Lewis. 1967. *The Pesticide Problem: An Economic Approach to Public Policy.* Resources for the Future, Washington, D.C.

Hendrick, R. D., F. L. Banner, T. R. Everett, and J. E. Fahey. 1966. Residue studies on aldrin and dieldrin in soils, water, and crawfish from rice fields having insecticide contamination. *J. Econ. Entomol.,* **59**: 1388-1397.

Hickey, J. J., and D. W. Anderson. 1968. Chlorinated hydrocarbons and eggshell changes in raptorial and fish-eating birds. *Science,* **162**: 271-273.

Holden, A. V., and K. Marsden. 1967. Organochlorine pesticides in seals and porpoises. *Nature,* **216**: 1274-1276.

Hunt, E. G. 1966. Biological magnification of pesticides. In: *Scientific Aspects of Pest Control.* Nat. Acad. Sci.-Nat. Res. Council, Publ. 1402, Washington, D.C., p. 251-262.

Hunt, E. G., and A. I. Bishoff. 1960. Inimical effects on wildlife of periodic DDD applications to Clear Lake. *Calif. Fish Game,* **46**: 91-106.

Hunter, C. G. 1968. *Allowable Body Burdens of Organochlorine Insecticides.* Proc. 5th Intl. Congr. Hyg. Prev. Med., Rome, October 8-12, p. 1-10.

Jukes, T. H. 1963. People and pesticides. *Amer. Sci.,* **51**: 355-361.

Kay, K. 1965. Recent advances in research on environmental toxicology of the agricultural occupations. *Amer. J. Publ. Health,* **55** (7) (suppl., part II): 1-9.

Keith, J. O. 1966. Insecticide contaminations in wetland habitats and their effects on fish-eating birds. *J. Appl. Ecol.,* **3** (suppl.): 71-85.

Kupfer, D. 1968. Effect of pesticides and related compounds on steroid metabolism. *Residue Rev.,* **19**: 11-30.

Maes, R., and A. Heyndrickx. 1966. Distribution of organic chlorinated insecticides in human tissues. *Meded. Rijskfac. Landbouwwetensch., Gent.,* **31**: (3): 1021-1025.

Matthysse, J. G. 1959. An evaluation of mist blowing and sanitation in Dutch elm disease control programs. N.Y. State Coll. Agr., Ithaca, N.Y., Cornell Misc. Bull. 30, 16 p.

Mitchell, E. L. 1966. Pesticides: properties and prognosis. In: "Organic Pesticides in the Environment." *Adv. Chem. Series,* **60**: 1-22.

Ratcliffe, D. A. 1967. Decrease in egg shell weight of certain birds of prey. *Nature,* **215**: 208-210.

Risebrough, R.W., D. B. Menzel, D. J. Marsten, Jr., and H. S. Olcott. 1967. DDT residues in Pacific sea birds; a persistent insecticide in marine food chains. *Nature,* **216**: 589-591.

Risebrough, R. W., P. Rieche, D. B. Peakall, S. G. Herman, and M. N. Kirven. 1968. Polychlorinated biphenyls in the global ecosystem. *Nature,* **220**(5172): 1098-1102.

Robinson, J. 1967. Dynamics of organochlorine insecticides in vertebrates and ecosystems. *Nature,* **215**(5096): 33-35.

———. 1969. The burden of chlorinated hydrocarbon pesticides in man. *Can. Med. Assoc. J.,* **100**: 180-191.

Robinson, J., A. Richardson, A. N. Crabtree, J. C. Coulson, and G. R. Potts. 1967. Organochlorine residues in marine organisms. *Nature,* **214**(5095): 1307-1311.

Rudd, R. L. 1964. *Pesticides and the Living Landscape.* University of Wisconsin Press, Madison, 320 p.

Sallis, J. D., and E. S. Holdsworth. 1962. Calcium metabolism in relation to vitamin D_3 and adrenal function in the chick. *Amer. J. Physiol.,* **203**: 506-12.

Sladen, W. J. L., C. M. Menzie, and W. L. Reichel. 1966. DDT residues in Adelie penguins and a crabeater seal from Antarctica. *Nature,* **210**(5037): 670-673.

U.S. Department of Agriculture. 1965. Press release, May, 1965.

———. 1967a. Agricultural Statistics.

———. 1967b. *Suggested Guide for the Use of Insecticides . . ."* Agriculture Handbook No. 331.

Walker, C. H., G. A. Hamilton, and R. B. Harrison. 1967. Organochlorine insecticide residues in wild birds in Britain. *J. Sci. Food Agr.,* **18**: 123-129.

Walley, W. W., D. E. Ferguson, and D. D. Culley. 1966. The toxicity, metabolism, and fate of DDT in certain icterid birds. *J. Miss. Acad. Sci.,* **12**: 281-300.

West, I. 1966. Biological effects of pesticides in the environment. In: "Organic Pesticides in the Environment." *Advan. Chem. Ser.,* **60**: 38-53.

———. 1968. Sequelae of poisoning from phosphate ester pesticides. *Ind. Med. Surg.,* **37**(7): 538, Abstract.

West, I., and T. H. Milby. 1965. Public health problems arising from the use of pesticides. *Residue Rev.,* **11**: 141-159.

Whitten, R. R., and R. U. Swingle. 1964. The Dutch elm disease and its control. USDA, Agr. Info. Bull. 193, 11 p.

Woodwell, G. M., C. F. Wurster, Jr., and P. A. Isaacson. 1967. DDT residues in an East coast estuary: a case of biological concentration of a persistent insecticide. *Science*, **156**: 821-824.

Wurster, D. H., C. F. Wurster, Jr., and W. N. Strickland. 1965. Bird mortality following DDT spray for Dutch elm disease. *Ecology*, **46**: 488-499.

Wurster, C. F., Jr., and D. B. Wingate. 1968. DDT residues and declining reproduction in the Bermuda petrel. *Science*, **159**: 979-981.

Pesticides: Eggshell Thinning and Lowered Production of Young in Prairie Falcons

J.H. Enderson and D.D. Berger

The precipitous decline of the peregrine falcon (*Falco peregrinus*) in both hemispheres has been attributed to certain agricultural toxic chemicals (Hickey, 1969), but North American prairie falcons (*F. mexicanus*) have not shown drastic population change. We report here on the relationships of experimentally introduced dieldrin and the natural contaminant DDE to aspects of nesting success of prairie falcons in Colorado and Wyoming in 1967 and 1968. The data show that high pesticide residues in egg contents, eggshell thinning, and pronounced hatching failure are correlated events.

We fed dieldrin to 43 of 78 nesting wild female prairie falcons under study by tethering dieldrin-contaminated starlings—fed 10 ppm dieldrin in their diet for 14 days—in sight of the perched falcons. Some females took as many as 12 starlings within 6 weeks prior to egg laying, and 23 took more than three. One egg per clutch was collected for pesticide analysis (dry weight), and the shells were measured, washed, dried, and weighed. The female was caught as soon as possible and a sample of subcutaneous fat taken by biopsy (Enderson and Berger, 1968) and analyzed (fat weight). Subsequently, the nesting success of the pair was determined. Tissue samples for organochlorine pesticide analysis were preserved by freezing until analyzed by WARF Institute, Inc., Madison, Wisconsin (Enderson and Berger, 1968; USFDA, 1963). Polychlorinated biphenyl compounds have been found in falcon tissue (Risebrough et al., 1968) and may have interfered with the determination of DDT and TDE in our samples; hence, we are not reporting these pesticides.

Nine dieldrin-treated starlings tested bore high whole-body levels of that pesticide and after 11 were fed, one per day, to each of three captive immature falcons we found high concentrations of dieldrin in the various falcon tissues (Table 1). All falcon samples analyzed bore DDE, sometimes in large amounts, as a nonexperimental contaminant, and we will distinguish between the effects of these compounds when possible.

Wild prairie falcons had small amounts of dieldrin in their adipose tissue relative to DDE (Table 1), a condition reported in peregrines (Cade et al., 1968; Enderson and Berger, 1968). Dieldrin levels in adipose tissues of wild experimental falcons fed more than three treated starlings averaged eight times those in untreated falcons and over half the mean DDE concentration.

In treated falcons, DDE plus dieldrin in fat averaged 237 ppm (range: 82-462). This probably represents substantially all pesticide residues present and is not far below the mean level of 392 ppm for these two pesticides, corrected to extractable fat, which we reported (Enderson and Berger, 1968) in adipose tissue from peregrines in western Canada. However, in the prairie falcons, dieldrin is relatively higher.

The relationship between dieldrin in the fat of females and that in their eggs appears to fit a straight line ($\hat{Y} = 0.21 + 0.17X$) and the cor-

TABLE 1. Dieldrin in prairie falcons fed contaminated starlings[a]

Item	n	Tissues and Residues (ppm)[b]			
		Dieldrin		DDE	
		adipose	brain	adipose	brain
Treated falcons[c]					
Bird 1	1				
before treatment	-	2.1	-	28.7	-
after treatment	-	694	3.04	168	0.4
Bird 2	1				
after treatment	-	370	2.8	27.5	0.3
Bird 3[d] (died)	1				
after treatment	-	-	11	-	1.9
Wild falcons (untreated)	21	10.5(S.E.:1.6)	-	169(S.E.:38)	-
Wild falcons (ate > 3 treated starlings)	20[e]	83(S.E.:13)	-	154(S.E.:22)	-

[a]Starlings contaminated in laboratory averaged 29 ppm dieldrin (whole body, wet basis), S.E. = 3.8, n = 9; two had traces of DDE

[b]Adipose tissue on fat basis; brain on wet basis

[c]Fed 11 dieldrin-contaminated starlings in 11 days

[d]The liver bore 29 and 4.6 ppm (wet basis) dieldrin and DDE, respectively, and muscle 17.5 and 3.8

[e]Samples not available for three others

relation coefficient is 0.87 ($n = 44$) when one anomalous case is omitted. Where dieldrin is below 25 ppm in fat, levels in eggs are below 8 ppm, and 60-100 ppm in female fat corresponds to 11-17 ppm in eggs.

DDE in female fat plotted against that in their eggs suggests a weak positive correlation. A mean of 24 ppm DDE was found in 70 eggs, or about 5 ppm on a wet basis at 80% egg moisture content.

In evaluating shell condition, Ratcliffe (1967) used a thickness index (weight/length × width). Eggs collected through 1940 (J. Hickey and D. Anderson, unpublished data) (Table 2) had a higher average index than those found in untreated birds in this study. Eggshells from 34 untreated birds had an average index significantly higher than eggshells from seven dieldrin-treated falcons where dieldrin exceeded 20 ppm in egg contents, averaging 41.5 ppm or 22 times the untreated group mean of 1.9 ppm. Only two eggs in seven clutches in the over-20 ppm group hatched. Eggs with less than 20 ppm dieldrin from treated birds had shells not significantly different from the untreated group. DDE varied widely among these samples.

The relationship of DDE in eggs compared to the shell index is shown in Figure 1 and includes only those eggs where DDE exceeds 80% of the total of all assayed residues. The grouped data show a significant decline in shell condition between some classes of residue levels and suggest a consistent decrease as residue levels increase.

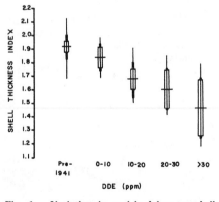

Fig. 1. Variation in prairie falcon eggshell thickness index (Ratcliffe, 1967) with changes in DDE (dry-weight basis) in egg contents where DDE exceeds 80% of total residues, pre-1941 data for comparison. Light line: range; heavy line: standard deviation; box: confidence limits on mean ($P < 0.05$); horizontal line: mean. Sample sizes are 32, 8, 13, 7, 7 (> 30 ppm group).

Dieldrin exceeded 30% of the total residues in 33 eggs, 32 of which were from dieldrin-treated pairs. The relationship of these data to shell condition is like that of DDE. A plot of DDE plus dieldrin against Ratcliffe's index for each eggshell shows the same general phenomena. Where the sum of the two residues is below 10 ppm, the mean index is 1.84 ($n = 10$); and where the sum is above 25 ppm, the mean is 1.57 ($n = 29$). The two means are statistically distinct (t-test, $P < 0.01$).

In eggs from 23 prairie falcons thought to have eaten more than three starlings, DDE plus dieldrin averaged 42 ppm (range: 5 - 145). This is about half the mean of DDE plus dieldrin we found in peregrine eggs in western Canada (Enderson and Berger, 1968) when the reported values were converted to a dry-weight basis (assuming 80% moisture).

Successful clutches (those hatching 3 or 4 eggs) had representative shell indexes averaging 1.77; those from clutches hatching none averaged 1.50. These means are statistically different (t-test, $P > 0.01$). Hatching occurred in only one of 10 clutches were representative-thickness indexes were less than 1.45, but in 22 instances where representative eggs had indexes above 1.75 only two sets failed to hatch. At 19 sites where no hatching occurred, fractured eggs were found at two sites and shell fragments below three others, with intact eggs remaining above. One broken egg was found at each of two sites where remaining eggs hatched.

We obtained shell measurements of representative eggs from 62 clutches where the production of flying young was determined (Table 3); lower thickness indexes accompany low fledgling success. Among these are 14 nest-sites where we found evidence that young were taken by people; in these cases we assumed that the young last seen at the site would have fledged.

Our data show that prairie falcons accumulate high levels of dieldrin in body fat after ingesting only a few highly contaminated prey items. Dieldrin in eggs from treated falcons was many times that found in eggs from untreated falcons, and a significant reduction in eggshell quality occurred. That a few prey carcasses can radically elevate residue levels in individuals may explain, in part, why residues of DDE acquired by wild birds varied greatly among our samples of body fat and eggs where falcons may be exposed to food varying in residue content.

Decreases in shell thickness in eggs from raptorial birds coinciding with environmental pollution by organochlorine pesticides (Ratcliffe, 1967) and an inverse relationship between eggshell thickness and levels of DDE in eggs from several regions (Hickey and Anderson, 1968) have been discovered. Our data show such a correlation for DDE in eggs from the same region. About 50-60 ppm DDE have been reported (Cade et al., 1968; Enderson and Berger, 1968) in eggs from northern peregrines; these levels coincide with very thin shells in prairie falcons (Fig. 1).

The hypothesis that thin shells result in lowered hatchability is plausible, and a causal rela-

TABLE 2. Prairie falcon eggs: Shell condition and pesticides

	n	Dieldrin (mean - S.E.)	DDE (mean - S.E.)	Ratcliffe's Index (mean - S.E., [range])
Representatives from clutches collected in Calif., Colo., Utah, So. Dak., 1909-1940 (J. Hickey and D. Anderson, unpublished data)	32	—	—	1.92[a] - 0.17 (1.68 - 2.13)
Representatives from clutches; untreated falcons in this study	34	1.9 - 0.4	27.5 - 9.5	1.65[a] - 0.03 (1.18 - 1.99)
Representatives bearing > 20 ppm dieldrin from treated falcons in this study[b]	7	41.5 - 4.2	47.1 - 8.1	1.50[a] - 0.04 (1.27 - 1.63)

[a] Significant $P < 0.05$.
[b] Mean of eggs with less than 20 ppm dieldrin not significantly different from untreated group.

TABLE 3. Production of young in prairie falcons compared to eggshell thickness

Thickness Index	n	Total No. yg. fledged	No. yg. fledged per pair	% of pairs fledging 1 or more yg.
> 1.75	21	54	2.6	75
1.45 - 1.75	31	51	1.6	50
< 1.45	10	3	0.3	10

tionship between the appearance of thin shells and increased egg breakage in British peregrine clutches has been suggested (Ratcliffe, 1967). Significant egg disappearance has been shown in captive sparrow hawks (*Falco sparverius*) where eggshell thinning was produced by low dietary intake of dieldrin and DDE (Porter and Wiemeyer, 1969). In the present study, we found evidence of egg breakage in 5 of 74 clutches, and the mean shell-thickness index of clutches where no eggs hatched and clutches where hatch was good are significant ($P < 0.01$).

These data establish a correlation between pesticide residues, thin eggshells, and poor hatching success. We predict prairie falcons in North America are involved in events leading to reduced reproduction and that northern peregrines, shown to bear high pesticide residue levels (Cade et al., 1968; Enderson and Berger, 1968; Enderson et al., 1968), will soon be found to have thin-shelled eggs.

Acknowledgments

This work was supported by NSF grant GB-5225, and Wayman Walker II provided invaluable field assistance.

References

Cade, T. J., C. M. White, and J. R. Haugh. 1968. Peregrines and pesticides in Alaska. *Condor,* **70**: 170.

Enderson, J. H., and D. D. Berger. 1968. Chlorinated hydrocarbon residues in peregrines and their prey species from northern Canada. *Condor,* **70**: 149.

Enderson, J. H., L. G. Swartz, and D. G. Roseneau. 1968. Nesting performance and pesticide residues in Alaskan and Yukon peregrines in 1967. *Auk,* **85**: 683.

Hickey, J. J. (ed.). 1969. *Peregrine Falcon Populations: Their Biology and Decline.* University of Wisconsin Press, Madison.

Hickey, J. J., and D. W. Anderson. 1968. Chlorinated hydrocarbons and eggshell changes in raptorial and fish-eating birds. *Science,* **162**: 271.

Porter, R. D., and S. N. Wiemeyer. 1969. Dieldrin and DDT: effects on sparrow hawk eggshells and reproduction. *Science,* **165**: 199.

Ratcliffe, D. A. 1967. Decrease in eggshell weight in certain birds of prey. *Nature,* **215**: 208.

Risebrough, R. W., P. Rieche, D. B. Peakall, S. G. Herman, and M. N. Kirven. 1968. Polychlorinated biphenyls in the global ecosystem. *Nature,* **220**: 1098.

U.S. Food and Drug Administration. 1963. *Pesticide Analytical Manual,* U.S. Dept. of Health, Education, and Welfare, FDA Adm. Publ. (revised 1964 and 1965).

JAMES H. ENDERSON
*Department of Biology
Colorado College
Colorado Springs, Colo. 80903*

DANIEL D. BERGER
*510 East MacArthur Road
Milwaukee, Wis. 53217*

Poisoning with DDT: Second- and Third-Year Reproductive Performance of *Artemia*

Daniel S. Grosch

In 1966, seven 3-liter population jars of brine shrimp were subjected to 1.00 ml doses of p,p' DDT from a dilution series dissolved in acetone. The objective was to assess alterations in fecundity and fertility. However, death caused five of the mass cultures to become extinct within 1 to 3 weeks. On the other hand, at the two sublethal doses, no induced genetic change was expressed in the F_1 generation (Grosch, 1967). Subsequently, in brother-by-sister matings, we have tested the reproductive performance of the 2nd and 4th generations. These data should reflect delayed dominant lethal phenomena as well as the segregation of deleterious recessive genes. Again, no genetic changes were demonstrable.

In both 1967 and 1968, five 3-liter population jars were available for comparison, because two subcultures had been instituted in 1966 by transferring more than 40 adult F_1 pairs from the treatment jars into jars containing our standardized brine (filtered seawater supplemented with 50 g of NaCl per liter). The jars compared were (1) a population subjected to 1×10^{-5} ppm DDT; (2) a subculture from no. 1; (3) a population subjected to 1×10^{-6} ppm DDT; (4) a subculture from no. 3; and (5) a control population which received only the acetone solvent. Subcultures 2 and 4 differ from the cultures in the original treatment jars 1 and 3 by providing an uncontaminated environment for the rinsed shrimp transferred thereto.

During the winters, these population jars were allowed to evaporate gradually, much like the process in many natural salterns. In early June, when I added 3 liters of distilled water to each jar, cultures reconstituted themselves. Each year in jars 1 and 3, despite an abundant emergence from cysts, the number of shrimp attaining maturity amounted to only about 60% of the number maturing in subcultures 2 and 4 and in acetone control 5. Nevertheless, when tested in individual quart jars, the reproductive performance of pairs was excellent (Table 1). By mid-July, when a second generation had contributed to the population, all five jars reached carrying capacity (250) and appeared identical. All cultures received 1 ml of yeast suspension daily, starting from the day of emergence of larvae.

The brother-by-sister matings of Table 1 reveal the excellent average reproductive performance of 15 early summer pairs in 1967 and 1968. The average number of broods and the number of zygotes per brood approach or surpass that of the acetone control. With standard errors ranging between 26 to 28, tabulated differences in the mean number of zygotes per brood are not statistically significant. By 1968, the percentages of zygotes encysted and of cyst emergence had returned to typical stock #3 values from levels interpreted as a response to environmental stress (Grosch, 1967). Also in 1968, survival to adulthood settled down to variation around 89%. For this measure of fitness, standard errors have ranged consistently between 6 and 8, so, even in 1967, little importance can be ascribed to a difference of only 11.7% between one of the treated populations (68.9%) and the acetone control (80.6%).

In addition to the data of Table 1, in both years the mean adult survival for both sexes of parents from the three groups exceeded 40 days and after the first brood the frequency of brood deposit was 3 to 4 days. These are characteristics of stock #3 from which the experimental populations were derived. Such results differ from the consequences of true mutagenic agents where pair matings have revealed inferiority of reproductive performance persisting over more than 20 generations (Grosch, 1966).

In June 1969, after overwinter evaporation, the entire contents of jars 1, 3, and 5 were analyzed by the Pesticides Residue Laboratory

TABLE 1. The reproductive performance of 15 pairs of second and fourth generation *Artemia* from populations treated with DDT in 1966

Population Treatment in 1966	Av. No. of Broods	Zygotes Encysted %	Cyst Emergence %	Zygotes per Brood	Survival to Adult %
Second Generation (1967)					
10^{-5} ppm	8.3	10.7	50.2	147.3	68.9
10^{-6} ppm	10.7	19.8	53.1	165.7	80.0
Acetone	8.5	23.3	50.9	121.3	80.6
Fourth Generation (1968)					
10^{-5} ppm	8.8	10.7	22.8	144.3	88.6
10^{-6} ppm	8.9	10.7	30.1	157.5	89.6
Acetone	9.4	18.0	19.4	145.3	89.3

Supported by PHS Grant ES-00044, Division of Environmental Engineering and Food Protection. Published as a short scientific report with the approval of the Director of Research, North Carolina Experiment Station.

of N.C.S.U.[1] The total p,p′ DDT content of jars 1 and 3 exceeded control background by 0.12 and 0.10 µg, respectively, and traces of DDD and DDE were present. (Limits of detection were 0.06 µg per sample). This is of the order of magnitude of doses delivered to individual insects to alter fecundity (Adkisson and Wells, 1962; Beard, 1965), but considerably less than dietary levels found lethal for adult crabs and oysters (Butler, 1969). Apparently, a fraction of our original dose has persisted as a toxic residue despite losses by (a) co-evaporation; (b) shrimp transfer for subculture and pair mating tests; and (c) conversion by yeast and other microorganisms (Kallman and Andrews, 1963; Miskus et al., 1965). Evidently, the persistent DDT took its toll of the swarm of nauplii when this was the predominant life form. Later in the summer, toxicity of the small amount of insecticide was minimized, presumably by its distribution in the system. This type of attrition is very difficult to detect in nature and we have been warned it may occur in the lower trophic levels of our salt marshes (Woodwell et al., 1967). Although nongenetic in character, the cyclic coexistence of pesticide residue and vulnerable stage in a life cycle can influence many generations.

References

Adkisson, P. L., and S. G. Wells. 1962. Effect of DDT poisoning on the longevity of the pink bollworm. *J. Econ. Entomol.*, **55**: 842-845.

Beard, R. L. 1965. Ovarian suppression by DDT and resistance in the house fly (*Musca domestica* L.) *Entomol. Exp. Appl.*, **8**: 193-204.

Butler, P. A. 1969. The significance of DDT residue in estuarine fauna. In: *Chemical Fallout/Current Research on Persistent Pesticides*, M. W. Miller & G. C. Berg (eds.), Charles C Thomas, Publisher, Springfield, Ill., Chap. 9.

Grosch, D. S. 1966. The reproductive capacity of *Artemia* subjected to successive contaminations with radiophosphorus. *Biol. Bull.*, **131**: 261-271.

———. 1967. Poisoning with DDT: Effect on reproductive performance of *Artemia*. *Science*, **155**: 592-593.

Kallman, B. J., and A. K. Andrews. 1963. Reductive dechlorination of DDT to DDD by yeast. *Science*, **141**: 1050-1051.

Miskus, R. P., D. P. Blair, and J. E. Casida. 1965. Conversion of DDT to DDD by bovine rumen fluid, lake water, and residue porphyrins. *J. Agr. Food Chem.*, **13**: 481-483.

Woodwell, G. M., C. F. Wurster Jr., and P. A. Isaacson. 1967. DDT residues in an East Coast estuary: A case of biological concentration of a persistent insecticide. *Science*, **156**: 821-824.

Daniel S. Grosch
North Carolina State University, Raleigh
(BioScience 20, no. 16, p. 913)

---

[1] Thanks are due Dr. T. J. Sheets, Director of the N.C. State Univ. Pesticides, Residue Research Lab. for the insecticide determinations.